T0330377

The Climate Resilient Organization

The Climate Resilient Organization

Adaptation and Resilience to Climate Change and Weather Extremes

Martina K. Linnenluecke

UQ Business School, The University of Queensland, Australia

Andrew Griffiths

UQ Business School, The University of Queensland, Australia

 Edward Elgar PUBLISHING

Cheltenham, UK • Northampton, MA, USA

Published by
Edward Elgar Publishing Ltd
The Lypiatts
15 Lansdown Road
Cheltenham
Glos GL50 2JA
UK

Edward Elgar Publishing Inc.
William Pratt House
9 Dewey Court
Northampton
Massachusetts 01060
USA

A catalogue record for this book
is available from the British Library

Library of Congress Control Number: 2014952142

This book is available electronically in the **Elgar**online
Social and Political Sciences subject collection
DOI 10.4337/9781782545835

ISBN 978 1 78254 582 8 (cased)
ISBN978 1 78254 583 5 (eBook)

Typeset by Columns Design XML Ltd, Reading
Printed and bound in Great Britain by T.J. International Ltd, Padstow

Contents

Figures

Tables

Introduction

On 20 March 2006, the Category 5 Tropical Cyclone Larry crossed the Queensland coast. Cyclone Larry caused extensive damage to the Wet Tropics regions in North Queensland, Australia. Areas highly affected included the regional towns of Tully, Innisfail and Babinda, as well as the Atherton Tablelands, with extensive flooding in the Gulf Country as the cyclone diminished into a tropical rainfall depression. Cyclone Larry was the most damaging cyclone to hit Queensland and had much greater impact than the Category 3 Cyclone Winifred that hit the same region in 1986. As the Far North Queensland economy relies heavily on both the tourism and primary industry sectors, the cyclone caused significant impacts on tourism operators and the primary industries of banana, sugar, beef, dairy, tree fruit, nut, lifestyle horticulture, forestry, pig, poultry, fishery, and aquaculture.

Initial estimates of lost gross value of agricultural production over the first 12-month period amounted to A$473 million, representing 50 per cent of the forecast agricultural output of the region. The likely range of the total financial impact was between A$365 and A$545 million. As Cyclone Larry wiped out an estimated 80–90 per cent of banana production, Australia experienced significant shortages in banana supplies for the remainder of 2006 and price hikes of 400 to 500 per cent across the country. Cyclone Larry also caused significant damage to both private and public infrastructure in the region. Older (pre-1986) residential and commercial properties suffered the greatest damage while damage to structures built after Cyclone Winifred was less severe. Damage to private infrastructure ranged from 15 per cent of homes at Flying Fish Point to 99 per cent at Silkwood. This damage in total inflicted pressure on the consumer price index (CPI). While individual extreme weather events cannot be directly attributed to climate change, their impacts paint a picture of damages that are projected to increase in the future as the climate changes.

This book looks at the issues arising from climate change and weather extremes for organizational decision-makers and policy-makers, and suggests that a future key activity will be to create climate change resilient organizations (including private sector firms and industries, as well as other types of organizations, such as not-for-profit organizations or local government). Even though public opinion is seemingly divided on the causes of and certainties around the occurrence of climate change, the scientific community has put forward a large and growing body of evidence which establishes that climate change is occurring, and that the resulting impacts are presenting very real, substantial, and tangible threats. Scientific data summarized by the Intergovernmental Panel on Climate Change (IPCC) shows that rising temperatures, changes in sea levels and melting glaciers are phenomena that are not occurring in a distant future, but outcomes of a changing climate that are already measurable in the present day (IPCC, 2007a; 2012; 2013). Impacts from climate change are expected to increase further in the future, particularly in vulnerable sectors and locations.

Even more concerning, it is expected that the greatest vulnerabilities of firms and industries, but also of settlement and society as a whole, will not occur due to gradual temperature increases alone (Wilbanks et al., 2007b). Projections show that the increasing surface temperature of the Earth will not only lead to gradual changes in mean climate conditions (for example, gradual increases in mean temperature), but also to larger climate variability and changes in the frequency, intensity, duration, spatial extent, and timing of extreme weather and climate events (IPCC, 2012; Meybeck et al., 2012). Impacts from climate change are expected to increase further in the future, particularly in vulnerable sectors and locations.

For industries and firms, the significance of any changes in the environment, particularly in the frequency or intensity in weather extremes, lies in their potential to bring about considerable adverse and unprecedented impacts on business activities (Hertin et al., 2003; Wilbanks et al., 2007b). These can occur in the form of direct impacts (direct exposure to damaging weather extremes) or can result from flow-on effects (indirect exposure, for example, from power outages, or impacts on important infrastructure) (Wilbanks et al., 2007b). The likely geographical distributions of such impacts, the timeframes over which they will occur, as well as the probabilities of the severity of impacts and particular future climate change scenarios are still surrounded by much uncertainty (Linnenluecke and Griffiths, 2011; Schneider et al., 2007; Wilbanks et al., 2007b). One issue that seems certain is that there are few areas and economies that will not be impacted upon by large-scale changes such as rising sea levels and coastal flooding, changes in patterns of extreme weather, as well as disruptions to agricultural systems (Brown and Funk, 2008; Smith et al., 2007).

Insurance statistics have demonstrated a greater likelihood, magnitude and diversity of damages from weather extremes over the past decade (Munich Re, 2012). Climate change and weather extremes have been cited as contributing factors, along with a number of other circumstances that have been converging and leading to this trend. Population growth and migration into higher risk areas such as coastal zones and cities and the industrialization of these areas are exacerbating the exposure of human settlements, industries and corporations to weather extremes (Munich Re, 2009). Other contributing factors are societal trends towards a greater reliance on tightly-coupled infrastructure (for example, financial systems, communications, energy and transportation) that often operate within narrow margins and have an associated heightened risk for large-scale failure, breakdown and outage when an adverse event occurs (Perrow, 1984; Rinaldi et al., 2001).

Different schools of thought have emerged with regard to responding to climate change. The safest way to avert dangerous consequences of climate change would be to take immediate, decisive and effective action to reduce greenhouse gas (GHG) emissions – this has been referred to as *mitigation* (or a *limitationist* view) (Kates, 2000). Yet, despite the obvious urgency of mitigation, progress on a global scale has been slow at best, and overall GHG emissions continue to rise. Consequently, it is becoming increasingly important to develop strategies that will enable society to survive and thrive in a climate-changed world alongside mitigation mechanisms. Such strategies, aimed at measures to reduce the exposure and vulnerability of natural and human systems to actual or expected climate change impacts are commonly referred to as *adaptation* (Dow et al., 2013; IPPC, 2007). However, anticipated changes in climate and weather patterns put great pressure on organizations, industries and society to not only adapt to climate change impacts but also to build *resilience*, that is, the capacity to absorb, withstand and recover from adverse impacts from climate change (for example, climate and weather extremes) that are beyond the circumstances that organizations, industries and society are adapted to cope with (Linnenluecke and Griffiths, 2010).

This book looks at both efforts towards mitigating climate change (Chapters 3–4) to help avoid 'dangerous anthropogenic interference with the climate system' as specified in Article 2 of the United Nations Framework Convention on Climate Change (UNFCCC), as well as processes of adaptation and methods to integrate resilience into social systems with a particular focus on private sector organizations (Chapters 4–7) – as a response strategy to those impacts of climate change that cannot be (or are not) avoided under the projected different scenarios of climate change (Denton et al., 2014). Together with efforts to implement sustainability, the integration of adaptation, mitigation and resilience can lead to climate-resilient pathways – through implementing strategies, choices and actions that reduce climate change and its impacts (Denton et al., 2014). Even though there is a conceptual distinction between adaptation and mitigation, both concepts are related to each other, and need to be treated as interrelated issues. Adaptation is a gradual process of adjustments undertaken over time to modify the coping range of an organization. A coping range is defined as those changes in climate-related variables (for example, warming or changes in precipitation) that an organization can tolerate without experiencing adverse consequences (Chapter 4). Resilience, on the other hand, is a characteristic of organizations that possess a sufficiently wide coping range and/or can quickly recover from situations that create vulnerability to their operations once the boundaries of the coping range have been exceeded (Linnenluecke and Griffiths, 2012). As noted by the 2007 IPCC *Synthesis Report*:

> There is high confidence that neither adaptation nor mitigation alone can avoid all climate change impacts. Adaptation is necessary both in the short term and longer term to address impacts resulting from the warming that would occur even for the lowest stabilization scenarios assessed … . Unmitigated climate change would, in the long term, be likely to exceed the capacity of natural, managed and human systems to adapt. Reliance on adaptation alone could eventually lead to a magnitude of climate change to which effective adaptation is not possible, or will only be available at very high social, environmental and economic costs. (Pachauri and Reisinger, 2007: 65)

More recently, the IPCC has concluded that even the most stringent GHG mitigation efforts cannot fully avoid further impacts of climate change over the next few decades, which makes adaptation inevitable (Denton et al., 2014; IPCC, 2013). Without mitigation efforts, however, a magnitude of climate change is likely to be reached that would make adaptation impossible for some natural systems, while for most human systems it would involve very high social and economic costs. Historically, studies such as those by Jared Diamond in his book *Collapse* indicate the fragility of social systems, and how rapidly they can break down under sustained environmental pressures.

In the organizational context, adaptation and resilience potentials are often context-specific and related to the characteristics of particular climate change impacts. While organizations may be able to undergo steady adaptations to gradual climate change (for example, gradual increases in mean temperatures), they might require resilience capacities to deal with disruptions that go beyond this gradual trend and are related to changes in extremes. For instance, consider an electricity utility company that may need to upgrade maintenance systems to deal with the impact of an increasing number of record-temperature days on its distributions systems. Such a company can adapt over the long run by introducing infrastructure upgrades to deal with warmer temperatures, however, it is also exposed to impacts that it has not adapted to in time (such as prolonged heat waves and excessive energy demand).

While a clear imperative for building adaptation and resilience to climate change exists, many essential issues still remain unresolved (Linnenluecke, 2013a). First, against which criteria or baseline data can adaptation progress and attempts at building resilience be evaluated and monitored? Second, who should bear the responsibility for making adaptation decisions and for building resilience, and what type(s) of climate change impact(s) should be prioritized? Third, who is supposed to carry costs, and who will benefit?

The public sector, especially local governments, will certainly have a major role to play in meeting adaptation needs (Linnenluecke, 2013a). Indeed, adaptation decisions, such as land use planning or disaster management, are commonly regarded as issues that should be dealt with by the public, rather than the private sector. Governments, however, are often slow to agree on changes and even slower to implement them. While the short duration of political planning and election cycles is certainly a contributing factor, the divided public opinion on climate change does also not create any strong impetus for policy action. In addition, the difficulty to reach political agreements at a global scale directly conflicts with the urgency of making decisions on how to address climate change.

The private sector, on the other hand, might react quickly and efficiently in implementing solutions. It can, therefore, be a valuable partner to support adaptation policies and strategies. Governments often expect the private sector to adapt, yet few incentives exist to promote coherent adaptations. Individuals or organizations (especially those in more exposed sectors, such as tourism or agriculture) might respond to environmental changes or market signals brought about by climate change, which is also referred to as 'autonomous adaptation' (European Commission, 2009). However, this type of adaptation is unlikely to be optimal from a societal perspective, as individual actors will pursue their own adaptation goals and not integrate their actions. There are constraints to autonomous adaptation such as uncertainty about climate impacts and the costs of implementing adaptation decisions (European Commission, 2009). Few business decisions are directly targeted towards achieving societal adaptation to climate change and in alignment with their communities' adaptation needs. Private companies are often hamstrung by short-planning timeframes, short market system rewards and the short-termism of chief executive officers (CEOs). It is noticeable that many corporations also have short life spans.

Overall, the private sector has not yet systematically considered the organizational implications of changes in the environment and the trends of weather extremes, for example, changes to the intensity and/or frequency of storms, floods and droughts. While some exposed companies, such as those in the reinsurance industry (for example, Munich Re, Swiss Re) have begun to undertake research into the risks associated with a changing climate and resulting impacts on their organizations, most current debates on climate change and the global corporate response are mainly focused on mitigation – that is, adjustments that organizations can take to reduce their GHG emissions, mostly in response to policy and legislative changes (Linnenluecke and Griffiths, 2010). The question of how organizations can cope with the physical impacts of global warming and more frequent and/or intense weather extremes has largely remained outside of these debates (Griffiths et al., 2012). A common argument is that the less effective mitigation efforts are, the more important adaptation will become. However, little is known about issues such as the synergies or trade-offs between adaptation/resilience, mitigation and possible other important objectives, for example, ecosystem protection (Kates et al., 2001). Little is also known about how these issues play out over different time scales in the future.

Advocating for greater attention to adaptation has not been without controversy. Critics have argued that a focus on adaptation can be interpreted as a tacit admission that mitigation efforts are no longer sufficient and/or worth pursuing (World Economic Forum, 2013). Critics have also argued that adaptation strategies might have unintended consequences and actually lead to maladaptive outcomes. For instance, a society that is prepared only for a narrow range of (anticipated) changes in climate might not be sufficiently resilient against the unexpected consequences of climate change, or changes in the frequency and intensity of extreme weather events (Linnenluecke et al., 2012). A well-adapted society might also be more complacent towards changing environmental conditions, as adverse impacts are not directly 'felt' and factored into decision-making. Furthermore, some adaptation actions may increase overall societal vulnerability rather than actually reduce it, and thus be maladaptive (European Commission, 2009). Examples of maladaptive outcomes are cooling technologies or water supply systems that increase energy consumption, or flood protection infrastructure that helps to reduce frequent, low-to-moderate magnitude losses, but does not protect against larger magnitude threats. Such defenses might also have additional negative side effects, for instance, they might disturb coastal and river systems and take away natural defenses (European Commission, 2009; New et al., 2011).

Given the observations and projections outlined above, the key aim of the book is to develop an understanding of how organizations can develop strategic responses to climate change and changes in the frequency and/or intensity of weather extremes. The frameworks presented here move beyond typical risk and crisis management perspectives. Since 2005, our Business School has offered a suite of education strategies on climate change and management. Over 100 corporate leaders completed a formal course, over 400 postgraduate students elected a Masters subject, and over 550 executives took part in two-day workshops. These courses showed that the key difficulty for executives lies not in accepting the reality of climate change once confronted with the scientific evidence, but in determining company-specific contributions to climate change and the impacts of climate change on their operations, in addition to the overall economic impacts of policy measures, for example, a price on carbon. Executives are faced with the challenge to not only determine their organization's carbon intensity and forge transition paths to alternative low-carbon business models – but also to consider adaptation and resilience measures. Most organizations are accustomed to deal with some climate variability such as seasonal changes or wet and dry seasons (Carter et al., 2007), but are usually challenged to accommodate conditions that occur with a much greater magnitude, frequency and/or rate of change (Linnenluecke et al., 2012; Wilbanks et al., 2007b). Determining strategic approaches to climate change is a complex task, especially in the construction, energy and infrastructure sectors.

Included in the book are several chapters that seek to provide a foundation for understanding, assessing and evaluating organizational responses to climate change and more frequent and/or intense weather extremes. The first chapters serve as an introduction to the topic, assess the risks (and potential opportunities) posed by climate change and outline how this topic has to date received little attention in both policy-making and organizational practice. We see a twofold role for policy developments and scientific research to support organizational engagement with climate change: organizational-level mitigation targets need to be clearly linked to global emission targets, and climate models need to inform organizational-level adaptation. Mitigation efforts require knowledge on how to establish verifiable and auditable emission inventories, and on how to implement abatement measures through changes in organizational policies, infrastructure and processes. Adaptation requires knowledge on how to evaluate future climate impacts on organizations on a relatively fine-grained geographic and sectoral scale (Linnenluecke et al., 2012; Wilbanks et al., 2007b). Executive responses will depend on a clear understanding of options for emission reductions and organizational transformation.

The remainder of the book looks at the concepts of organizational adaptation and resilience to both impacts from gradual climate change and changes that occur with greater magnitude, frequency and/or rate of change than expected. We cover methodological and assessment issues – for instance, innovative methods and pathways to study adaptation needs – and the development of organizational resilience potentials under different climate change scenarios. The book thereby directly addresses issues associated with uncertainties about future climate change outcomes across temporal and spatial scales. The book also seeks to provide insights into what leads to the resilience of organizations, industry or society, and into the variables that should be considered in a decision-making context (Linnenluecke and Griffiths, 2012). Last, we also discuss the potential inability of organizations to adjust to changes in climate and weather extremes, and implications in terms of a necessity of a geographical shift of organizational and industrial activities. The key contribution of the book is that it extends existing debates on mitigation and addresses adaptation and resilience, and the connection between these concepts – and provides practical tools and frameworks that come from empirical evidence. The book seeks to close gaps in our understanding of organizational challenges associated with a changing climate.

Structure of this Book

This book is based on the premise that there is a twofold role for policy developments and scientific research to support organizational adaptation, mitigation and resilience to climate change impacts. First, organizational-level mitigation targets need to be clearly linked to global emission targets to avoid the worst consequences of climate change. At the same time, and given that not all adverse impacts of climate change can be mitigated, even with significant emissions cuts, models and projections about future climate impacts are required to inform societal and organizational-level adaptation targets. The most appreciable positive results can be made when integrating adaptation decisions into large-scale infrastructure developments and long-term planning frameworks, especially in regions that are rapidly developing and/or are highly vulnerable to climate impacts.

The book's aims are as follows:

- Explore organizational issues resulting from global environmental change, especially climate change.
- Provide an understanding of climate policy developments and their effectiveness.
- Develop insights into effective adaptation, resilience and response strategies of organizations.
- Discuss the innovative responses of organizations and the development of climate-resilient pathways as an impetus to promote change.

This book sets out to address these aims in Parts I and II. Part I starts by analyzing the impacts brought about by the aggregate levels of industrial activities, and outlines how the local manifestations of global changes impact organizations and industries. We also turn our attention to the policy-level of analysis and international progress on adaptation policy. Part II translates scientific knowledge on global climate change processes into information that has relevance to local decision-makers, recognizing that effective solutions require the involvement of actors across multiple scales and levels.

PART I

The changed environment

1. Organizations and global environmental change

> Not all organizations adapt equally well to the environment within which they grow.
>
> Many, like the dinosaur of great size but little brain, remain unchanged in a changing world.
> **Charles Handy**

INTRODUCTION

Since the beginning of the Industrial Revolution in the middle of the eighteenth century, global concentrations of greenhouse gases (GHGs) such as carbon dioxide (CO_2), methane (CH_4), and nitrous oxide (N_2O) have risen significantly and have substantially contributed to the warming of the Earth's atmosphere through the so-called greenhouse effect. The increase in CO_2 concentrations can primarily be attributed to the burning of fossil fuels such as coal, natural gas, and oil, as well as land-use changes such as deforestation. Increases in methane and nitrous oxide have primarily been linked to agriculture expansion and intensification (CSIRO, 2009). Examples include methane emissions from livestock and rice cultivation, or emissions of nitrous oxide from fertilized soils (Foley et al., 2011). Even though the extent to which climate change will impact society is still debated, it is clear that the role of business activity is a key matter in the debate about climate change. Businesses have been central to the creation of the wealth and technologies that have transformed society (Michaelis, 2003), but, at the same time, economic growth has also been a major driver behind the intensification of natural resource use and consumption to unsustainable levels.

Overall, there are very few aspects of current society that are not organized into or impacted upon by corporate activities. Consolidating and systemizing human production efforts into organizations has allowed for the accomplishment of economic activities beyond the means of individuals. This was achieved through the division of labor into various tasks to be performed and the introduction of organizational structures to allow for a coordination of these tasks (Mintzberg, 1983). Organizations have taken important roles in society in providing services, ranging from the provision of energy to transportation, food and health services. However, the industrialization and the arrival of the modern corporation at the turn of the twentieth century have led to concerns about the impacts of rational, utility- and profit-maximizing behaviors on society and the environment. It soon became clear that the growing levels of economic growth and industrialization and the primary pursuit of share-holder interests led to the exploitation of finite flows of natural resources, rather than their long-term conservation for the benefit of all (Nordhaus and Radetzki, 1994; Ostrom, 1999). The limits of exponential growth models and their impacts on natural systems became visible as they did not take into consideration the limitations of natural systems.

The available scientific evidence makes it clear that economic growth is not just interfering with the Earth's climate system. The scale of human activity, together with population growth and growing levels of resource consumption, has resulted in important changes to the Earth's systems. In addition to changes within the Earth's climate system, these include: accelerated biodiversity loss, interference with the nitrogen and phosphorus cycles, stratospheric ozone depletion, ocean acidification, depletion of global freshwater resources, adverse changes in land use, rising levels of chemical pollution, and atmospheric aerosol loading (Bateman et al., 2011; Rockström et al., 2009b). Studies assessing the state of the environment report that many ecosystems are in an alarming state of decline, and a further significant degradation is projected (Millennium Ecosystem Assessment, 2005b). The United Nations High-Level Panel on Global Sustainability concluded that 'The current global development model is unsustainable. We can no longer assume that our collective actions will not trigger tipping points as environmental thresholds are breached, risking irreversible damage to both ecosystems and human communities' (United Nations, 2012: 4).

Indeed, the scale of human and industrial activities is now seen as severe enough to potentially bring about large-scale, adverse global change. The US Global Change Research Act 1990 defines 'global change' as 'changes in the global environment (including alterations in climate, land productivity, oceans or other water resources, as well as atmospheric chemistry and ecological systems) that may alter the cap-acity of the Earth to sustain life'. This definition is directly expressing concerns about the large-scale degradation of the Earth's carrying cap-acity and the far-reaching impact of modern economic activities on the biosphere, natural resource depletion and ecosystem functioning.

Despite the large number of environmental problems caused by human activity, climate change has emerged as one of the most pressing issues, in particular due to concerns about changes in weather and climate patterns, as well as concerns about a greater occurrence of weather extremes and resulting disasters. It is estimated that humanity has by now transgressed the boundary level of CO_2 emissions that should not be transgressed if we are to avoid unacceptable global environmental change (Rockström et al., 2009a). Many impacts of climate change are already observable. Furthermore, climate change is an important driver and accelerator of vulnerabilities in other systems, such as agricultural systems and freshwater resources (IPCC, 2007a, 2013).

The global reach of corporations and the role of business activities in bringing about large-scale environmental change have triggered political and societal discussions about the implementation of more sustainable business practices. However, despite a growing number of initiatives (some with considerable success at lowering resources and energy consumption), environmental degradation and GHG emissions levels are overall on the rise. Researchers have noted the paradox surrounding corporate sustainability. On the one hand, actions such as the implementation of energy-saving measures have been among the most popular topics within the business community. Recent years have also seen a significant uptake of environmental reporting under the mantra of: 'what gets measured gets acted upon' (Confino, 2013). Sustainability is therefore no longer a fringe topic and organizations have started to act on ecological issues. On the other hand, collective action on climate change is not strong enough to prevent global emissions from rising, and to avoid an overall worsening of the state of the environment (Whiteman et al., 2013). Some contributions aside, action on climate change is a topic that has commanded little interest in business practice, management journals and business schools.

Looking at GHG emissions data, some of the world's largest companies are now emitting more GHGs than entire nations. The emissions of large, resource-intensive corporations such as Exxon Mobil are larger than those of smaller European states such as Belgium (see Figure 1.1), not measuring other negative externalities such as environmental pollution and ecosystem degradation. The extent of an individual corporation's contribution to climate change and other types of adverse global changes is still a topic absent from many boardrooms – in some cases because the impacts of corporate activities on the environment have not been fully assessed, in other cases because climate change is not perceived as an issue that requires immediate management attention beyond compliance with regulatory requirements. This is not to say that individual corporate

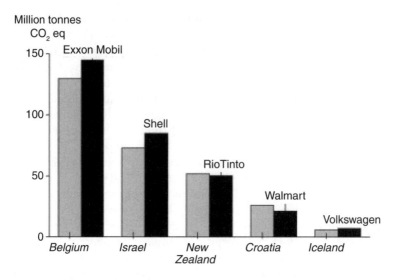

Source: Patenaude (2011). Reprinted from *Global Environmental Change*, Vol. 21, Iss. 1, Patenaude, G. (2011), 'Climate change diffusion: While the world tips, business schools lag', pp. 259–71. Copyright (2011) with permission from Elsevier.

Figure 1.1   Carbon emissions by companies

actions have been unsuccessful in addressing sustainability or mitigating emissions; rather, progress on an aggregate level is slow at best. In addition, little is known about the extent to which attempts at integrating sustainability into corporate decision-making actually contribute to eco-logically beneficial outcomes (Kallio and Nordberg, 2006). The potential issue is that by making an organization more efficient, we also provide this organization with the capacity to use more resources in an efficient and unsustainable way.

The lack of stringent reductions in global GHG emissions and the resulting greater potential for adverse consequences of climate change has resulted in researchers putting forward stronger arguments for creating adaptation and resilience to climate change impacts in addition to a focus on mitigation (Dow et al., 2013). At the 2013 First European Climate Change Adaptation Conference in Hamburg, Germany, Nigel Arnell (Director of the Walker Institute for Climate Change Research and Professor of Climate System Science in the Department of Meteorology at the University of Reading) concluded in his address to the conference that mitigation might reduce some, but does not eliminate the impacts of climate change. Overall, given that the consequences of climate change can increasingly not be avoided as climate change progresses, a strong imperative for adaptation and resilience strategies to address projected climate impacts and related changes in ecosystems thus emerges. The feasibility of any action will thereby largely depend on the future state of society and the economy (Arnell, 2013).

A Brief History of Organizations and the Environment

While many decision-makers are aware of the topic of sustainability and have started to consider changes to their organizations' production processes and service delivery to lower their resource inputs and through-puts, organizational behavior is largely occurring in a way that is decoupled from a changing environment. The most frequently cited reasons for a lack of more stringent business action on climate change are short-termism and profit-orientation. This leads to a tendency of organizational decision-makers to rarely consider the long-term impli-cations of organizational actions on the environment as well as the long-term impacts of a changed natural environment on their organ-izations' operations (Winn et al., 2011). However, the possibility that unsustainable levels of resource consumption can result in adverse environmental changes with significant negative outcomes for society (not only on a local but also on a global scale) is fully recognized and was already extensively documented by the nineteenth century (Malthus, 1878; Marsh, 1864; Mill, 1848).[1]

The warming effect of gases in the atmosphere was first discovered by the French mathematician Jean-Baptiste Fourier in 1827. Fourier pointed to the similarities between the earth's atmosphere and the glass of a greenhouse, which later led to the term 'greenhouse effect'. Further early research was undertaken in 1860 by John Tyndall who measured the absorption of infrared radiation by water vapor and CO_2, as well as by Svante Arrhenius in 1896 who calculated the effects of an increasing concentration of GHGs in the atmosphere (Houghton, 2004). From the beginning of the industrial revolution onwards, it became increasingly evident that the activities of modern corporations were becoming a major contributor to environmental problems and increasing levels of GHGs in

---

[1] Knowledge about the environmental impacts caused by human activity dates back much further, including traditional knowledge about how deforesta-tion, and inappropriate resource use such as water or soil depletion led to the decline and collapse of ancient civilizations. See Weiskel, T.C. (1989), 'The ecological lessons of the past: an anthropology of environmental decline', *The Ecologist*, **19**(3): 98–103.

the atmosphere. Environmental problems and concentrations of GHGs were intensifying with the aggregate level of activities and resource exploration.

Despite these initial studies, research on global warming came to a halt until the 1950s when scientists realized that several regions of the world had warmed significantly during the previous half-century. The first expression of concern about increasing levels of GHG emissions and their contribution to climate change was made by Revelle and Suess (1957) who published a paper on the build-up of CO_2 in the atmosphere. In the same year, regular measurements of the level of CO_2 in the atmosphere were started in the US from the observatory of Mauna Kea in Hawaii (Houghton, 2004).

A systematic identification of the role of industries and organizations in driving environmental change did not begin until the 1960s and 1970s, with an initial focus on environmental pollution and degradation brought about by corporate activities. At the time, highly visible ecological problems and growing awareness about the state of the environment galvanized public attention to oppose the negative health and environmental effects caused by organizations, and created a wave of support for stricter government regulations (Fischer and Schot, 1993; Hart, 1995, 1997). The book *Silent Spring* by Rachel Carson (1962) has been cited by many as a turning point in raising societal awareness about the severity of environmental degradation, and became a catalyst for the environmental movement of the 1960s and 1970s. In her book, Carson presented findings that agricultural pesticides (in particular DDT) were building up to dangerous levels and causing damage to both animal species and to human health. While the industry journal *Chemical Week* initially denounced Carson's findings, more and more evidence emerged which underpinned her claims and pointed to serious environmental problems brought about by industrial activities. These included the dispersion of toxic chemicals and waste (Maugh, 1979), the extinction of species (Bishop, 1978), acid rain (Likens and Bormann, 1974) and issues such as widespread air pollution, water pollution and soil contamination (Nriagu and Pacyna, 1988). The US Science Advisory Committee, tasked by John F. Kennedy to review the claims presented in *Silent Spring*, largely confirmed Carson's conclusions.

The media coverage following *Silent Spring*, along with a number of high-profile environmental disasters (for example, mass deaths of fish, major oil spills such as the Amoco Cadiz oil spill, the Love Canal scandal in the US, as well as mercury and cadmium poisoning in Japan) fueled a public outcry over environmental pollution and degradation caused by organizations. The First Earth Day was held in the US, not only capitalizing on the emerging environmental consciousness, but also on the growing awareness of social issues and the anti-war protest movement in the country. Although the environmental movement paid attention to the cumulative effects of individuals' consumption practices along with rapid population growth and economic development, especially in developing regions (Ehrlich, 1968), organizations were generally identified as the major culprits for environmental problems as a result of their bigger reach, greater resource access, and greater economic interests compared to individuals (Hart, 1997). Yet, despite growing public pressure and new government regulations during the 1970s and early 1980s, organizations did in most cases little more than to comply with legal requirements, and often attempted to fight or hinder them (Walley and Whitehead, 1994). Organizations needed to establish and finance environmental compliance functions alongside existing compliance functions such as health and safety, thus increasing their costs.

The extent of environmental degradation was eventually recognized on an intergovernmental level and led to the 1972 United Nations Conference on the Human Environment (UNCHE) in Stockholm, Sweden. The UNCHE was a first attempt to quantify the global human impact on the environment and to define principles for addressing the challenge of environmental preservation. Following this conference, the United Nations Environmental Programme (UNEP) was established. At the same time, the publication of *Limits of Growth* by the Club of Rome (Meadows et al., 1972) sparked a debate around the compatibility of economic and population growth with environmental protection. The 'growth versus environment' debate that followed questioned the validity of growth as a goal of society in general, assuming that indefinite future growth is impossible (Kidd, 1992). In order to prevent further environmental destruction, some economists proposed a shift towards a 'steady state economy' with no further growth (Daly, 1973, 1974) or even a reduction in both economic activity and population levels (Grant, 2000). Such propositions were viewed by many as too radical, thus shifting the discussion away from a 'growth versus environment' debate (Pearce et al., 1989) towards notions that some form of sustainable development is possible. The 1987 report *Our Common Future* by the World Commission on Environment and Development (WCED), for instance, popularized the notion of 'sustainable development', defined as: 'development which meets the needs of the present without compromising the ability of future generations to meet their own needs' (WCED, 1987).

During the mid- to late-1980s, and among further environmental disasters (for example, the Exxon Valdez oil spill, the Chernobyl nuclear disaster and the Bhopal gas disaster), several leading organizations began to change their position from ignoring or resisting environmental pressures to attempts to profit from them (Fischer and Schot, 1993; Starik and Marcus, 2000). Their responses were referred to as 'market-driven environmentalism' (Post and Altman, 1994: 65) and led to a widespread introduction of environmental management systems (for example, ISO14001) along with written policy statements, larger environmental departments and more extensive external contacts with a range of stakeholders including environmental groups (Fischer and Schot, 1993; Starik and Marcus, 2000). Organizations were able to achieve easy, but often very significant improvements and cost-savings, mainly through the introduction of pollution prevention and waste minimization initiatives (Post and Altman, 1994; Walley and Whitehead, 1994). Consequently, environmental considerations also started to play a greater role in capital investment decision-making and in new product development (Starik and Marcus, 2000).

Organizations attempted to capitalize further on the cost and competitive opportunities of engaging with environmental issues. Many organizations started to analyze the impact of their engagement with environmental issues on their corporate profitability and competitive advantage, for instance through environmental impact or life cycle assessments. Some organizations sought to develop internal capabilities to achieve a sustainable competitive advantage, including measures aimed at preventing pollution and improving the efficiencies of resource use (so-called eco-efficiencies) (Hart, 1995, 1997; Lovins et al., 1999). During this time, organizations also faced increasing pressures to disclose environmental information, which gave rise to environmental performance reporting. A wide variety of terms have been used to describe this development, including corporate 'greening' (Gladwin, 1993) and 'corporate environmentalism' (Hoffman, 2001; Jones and Levy, 2007). However, eco-initiatives and efficiency improvements did not automatically translate into total environmental performance improvements that can be attributed to overall consumption growth (Schot et al., 1997; Starik, 1995). The pursuit of sustainability was driven by the primacy of economic considerations over environmental concern for many, if not most corporations.

The late 1980s and early 1990s saw the emergence of a different type of environmental concern beyond environmental protection *per se*. The environmental movement of the 1960s and 1970s was driven by media attention and stakeholder pressures in response to highly visible, yet local and often reversible forms of pollution and environmental degradation. However, the discovery of the 'ozone hole' and the realization that climate change was posing a longer-term, irreversible, and global threat, were primarily driven by scientific research. Research findings started to point to environmental problems at a much larger scale (Bodansky, 2001). In 1988, the World Meteorological Organization (WMO) and the UNEP established the Intergovernmental Panel on Climate Change (IPCC) in an attempt to bring together international climate experts to synthesize the most recent climate science findings, and assess the state of scientific knowledge on climate change. These developments (summarized in Table 1.1) culminated in the 1992 United Nations Conference on Environment and Development (UNCED) in Rio de Janeiro and the United Nations Framework Convention on Climate Change (UNFCCC) which had great impact on the international policy agenda (Cantley-Smith, 2010), and resulted in the adoption of climate policies on a national level (as outlined in Chapter 4).

Increasing concerns about declining environmental quality, increasing population growth, rising living standards, and the increasing rates of consumption of resources (for example, of energy, natural resources and food) also led to an increasing level of environmental legislation. During the 1960s and 1970s several environmental laws were passed that introduced water and air quality standards in highly industrialized countries, including the US and countries across Europe (for example, the Clean Air Act 1963, the Clean Air Act 1970, the Water Quality Act 1965, the Air Quality Act 1967, the Safe Drinking Water Act 1974, and the National Environmental Policy 1969 in the US). This also led to the formation of corresponding institutions such as the US Environmental Protection Agency (US EPA). Increasingly, regulations were introduced relating to consumer and environmental protection (for example, the US Consumer Product Safety Act 1972) as well as the management of hazardous materials (for example, the US Toxic Substances Control Act 1976). In addition, greater attention was paid to waste management, resource efficiencies and the regulation of agricultural and industrial practices.

At around the same time, the notion was advanced that businesses can and should help society in achieving the three goals of environmental protection, social equity and economic prosperity (Elkington, 1997). The World Business Council on Sustainable Development (WBCSD) coined the term 'eco-efficiency' in the 1992 publication *Changing Course* to refer to what the WBCSD saw as a new management philosophy to encourage businesses to implement environmental improvements that result in parallel economic benefits (Schmidheiny, 1992). However, the wide acceptance of achieving 'eco-efficiencies' as the key strategic theme for businesses in relation to achieving sustainable commitments and activities has not been without criticism. For one, issues concerning

Table 1.1   *Main events preceding international policy action on climate change*

Date	Event	Implications		
			Increased awareness of environmental issues	
				Increased awareness of social/human rights issues
1896	Arrhenius develops calculations that show the effect of an increasing concentration of GHGs	The calculations provide a scientific basis for understanding the greenhouse effect		
1957	Revelle and Suess publish on the rise of carbon dioxide (CO$_2$) in the atmosphere (Revelle and Suess, 1957)	Forerunner of the climate change debate; ongoing measurements of atmospheric CO$_2$ concentrations begin in Hawaii		
1962–8	The book *Silent Spring* (Carson, 1962), the article *The Tragedy of the Commons* (Hardin, 1968), and the book *The Population Bomb* (Ehrlich, 1968) are published	Emergence of the early Environmentalist Movement as the publications and highly visible environmental problems lead to increased public awareness of environmental issues		
1970	First Earth Day	Beginning of the mainstream Environmentalist Movement		
1972	The UN Conference on the Human Environment is held in Stockholm and leads to the establishment of the United Nations Environment Program (UNEP) as well as several national environmental protection agencies	Environmental degradation is recognized as a serious threat to development, particularly to those living in absolute poverty		
1972	The Club of Rome publishes *Limits to Growth* (Meadows et al., 1972)	Emergence of *no growth/slow growth philosophy*, advocating no further economic growth (Daly, 1973, 1974, 1993) or even a reduction in both economic activity and population limits		
1973	OPEC oil crisis			
1973	Daly publishes *Toward a steady state economy* (Daly, 1973)			
Late 1970s	Views emerge that alternatives to economic growth exists, and that some form of sustainable development is possible	Concepts such as 'eco-development', or 'alternative development' supersede the no growth/slow growth philosophy		
1985	Meeting of the World Meteorological Society, UNEP and the International Council of Scientific Unions on the issue of climate change	Climate change moves up the political agenda		

1987	The World Commission on Environment and Development (WCED, founded in 1983 by the UN) published the report *Our Common Future*	The WCED report states that: 'sustainable development is development that meets the needs of the present without compromising the ability of future generations to meet their own needs' (WCED, 1987: 43)
1988	The Intergovernmental Panel on Climate Change (IPCC) is established to review and compile the existing body of research on climate change at the time	The IPCC provides a first comprehensive assessment of global warming and a broad range of climate change-related topics; the first IPCC report serves as basis for negotiating the United Nations Framework Convention on Climate Change (UNFCCC)
1992	The United Nations Conference on Environment and Development (UNCED, or The Earth Summit) is held in Rio de Janeiro, outcomes are the 'Agenda 21' (an action plan to achieve worldwide sustainable development adopted by more than 178 governments), the Rio Declaration, the Statement of Forest Principles, the UNFCCC, and the UN Convention on Biological Diversity	The UNCED led to aspirational plans urging countries to renew their commitment to sustainable development. Plans refer to 'sustainable development' and 'sustainability', however the terms are not fully defined or explained (United Nations, 1992)
1997	Adoption of the Kyoto Protocol (an agreement made under the UNFCCC) by most industrialized nations (United Nations, 1997)	Parties to the Kyoto Protocol are required to achieve 'sustainable development' via the reduction of GHG emissions
1997	Elkington publishes *Cannibals with Forks* (Elkington, 1997)	Elkington introduces the Triple Bottom Line concept for measuring organizational success, this includes the measurement of environmental and social performance besides economic success
1997	The Global Reporting Initiative (GRI) launched which sets a framework for reporting on economic, environmental, and social performance	
1999	Establishment of the Dow Jones Sustainability Index (DJSI) to track the performance of the global sustainability leaders	The definition by the DJSI states that: 'corporate sustainability is a business approach that creates long-term shareholder value by embracing opportunities and managing risks deriving from economic, environmental and social developments (as cited in Rooney, 2007: 138)'
2000	United Nations Millennium Summit	Agreement on the Millennium Development Goals (United Nations, 2000) by world leaders; goals are set to combat poverty, hunger, disease, illiteracy, environmental degradation and discrimination against women

Date	Event	Implications
2002	World Summit on Sustainable Development held in Johannesburg	The World Summit suggests 'partnerships' between community organizations and businesses as an approach to achieve sustainable development
2005	The Kyoto Protocol enters into force on 16 February 2005	The Kyoto Protocol sets the binding goal to reduce GHG emissions by at least 5 per cent from 1990 levels in the commitment period 2008–12
2006	*Millennium Ecosystem Assessment Report* is released (Millennium Ecosystem Assessment, 2005a)	The Report outlines the need for 'sustainable development' and 'sustainable resource use'
2006	Publication of the Stern Review (Stern, 2007; Stern et al., 2006); release of the IPCC *Fourth Assessment Report* (Pachauri and Reisinger, 2007)	The *Stern Report* and *IPCC Fourth Assessment Report* confirm the scientific basis for climate change and argue for the urgent need for adaptation and mitigation strategies
2008–12	First commitment period of the Kyoto Protocol	Parties with commitments under the Kyoto Protocol (38 industrialized countries and the European Community) commit to reduce their aggregate CO_2e emissions: on average by 5 per cent against 1990 levels
2012 on-wards	Successor of Kyoto Protocol	Second round of binding GHG emission targets for Europe, Australia and some other developed countries

Sources: IISD (2006), Griffiths and Limmenluecke (2008), Kidd (1992), US Senate (1997), and as indicated in timeline.

equity and other social properties are not included in the concept. In addition, computing 'environment improvements' in relation to economic benefit faces additional issues: what type of improvements should be included in any calculation, what boundaries are appropriate, and what metrics should be used (Ehrenfeld, 2005)?

The pursuit of eco-efficiencies has also been criticized for not leading to any significant radical changes in business models, away from unsustainable models of production and consumption. For instance, rather than 'revolutionizing' their business models, many companies responded to the slowly emerging climate policy landscape by making operational improvements and eliminating inefficiencies in energy- or fuel-intensive processes that stem from poor planning and the inefficient use of natural resources. In many cases such strategies allowed companies to reduce carbon emissions while also exploiting benefits associated with lower resources use (Hart, 1995, 1997). Well laid-out energy or resource efficiency programs have certainly significant benefits not only for reducing emissions, but also for improving financial performance and competitive advantage in the short term if cost-savings were being realized (Sharma and Aragón-Correa, 2005). However, it needs to be questioned whether established patterns of organizational responses with a focus on complying with regulation and seeking to exploit efficiencies are suited to respond to a vastly changed external environment in a new era of environmental change.

Organizations that only pursue efficiency gains and emission reduction strategies may neglect building adaptive capacity to the physical impacts that climate change will have on their operations (along with social, political and economic impacts that will emerge in the longer term). Indeed, many managers have not yet systematically considered the industry-level and organizational-level implications of a more volatile natural environment, such as changes in the intensity and frequency of storms, floods, and droughts. While some organizations (for example, agricultural companies) have traditionally been more exposed to weather extremes and have built adaptive responses, most current debates on climate change and corporate responses are limited to reducing GHG emissions. Exceptions can be found in sectors that are strategically affected by climate change and have begun to investigate the impacts of future trends of weather extremes on their business (for example, reinsurance firms such as Munich Re and Swiss Re).

Is There a Tipping Point?

Society may be lulled into a false sense of security by presuming that adaptation is possible due to the often smooth and gradual projections of climate change, and that that climate change thus does not pose any real threat (Lenton and Schellnhuber, 2007). Indeed, larger-scale changes beyond gradual temperature rise are often not considered in the public debate, or are seen as highly unlikely. However, scientists have put forward evidence concerning the existence of so-called tipping points in relation to climate change and resulting impacts (Kriegler et al., 2009; Nepstad et el., 2008). A *tipping point* can be defined as a threshold beyond which a system moves into a qualitatively different state, usually brought about by a small level of additional change or pressure (Lenton et al., 2008). An example can be a light switch – if only a small amount of pressure is applied, the light will not turn on. However, beyond a certain amount of pressure, the light switch will flick into a different state – and turn the light on (Gunderson and Holling, 2002). Exceeding a tipping point might lead to what is referred to as 'dangerous' climate change. The 1992 UNFCCC text (see Chapter 3) states that:

> [t]he ultimate objective of this Convention and any related legal instruments that the Conference of the Parties may adopt is to achieve, in accordance with the relevant provisions of the Convention, stabilization of greenhouse gas concentrations in the atmosphere at a level that would prevent dangerous anthropogenic interference with the climate system. Such a level should be achieved within a time frame sufficient to allow ecosystems to adapt naturally to climate change, to ensure that food production is not threatened and to enable economic development to proceed in a sustainable manner.[2]

The objective 'to achieve [...] stabilization of greenhouse gas concentrations in the atmosphere at a level that would prevent dangerous anthropogenic interference with the climate system' has been criticized as being too vague, as the meaning of 'dangerous anthropogenic interference' can be interpreted in many different ways. This statement does not specify a level of global warming that is dangerous (New et al., 2011), nor does it provide any specific timeframe and/or targets for actions. While this is problematic for the implementation of any GHG reduction targets, it is not surprising that the statement by the UNFCCC has not been more specific regarding what constitutes 'dangerous' levels of climate change. The existence of tipping points in relation to climate change (and the

---

[2]   The full text of the Convention is available, accessed 20 October 2014 at http://unfccc.int/resource/docs/convkp/conveng.pdf.

quantification of accumulated change needed to reach a tipping point) is still surrounded by uncertainty. Nonetheless, there is ample evidence on a smaller scale that human activities can push components of the Earth's systems (for example, ecosystems) past a critical threshold into qualitatively different modes, leading to the assumption that such changes can also occur on a planetary scale (Lenton et al., 2008). Examples of possible consequences of threshold exceedance include the potential collapse of the Atlantic thermohaline circulation (THC), dieback of the Amazon rainforest, and the collapse/decay of major ice sheets (Kriegler et al., 2009; Nepstad et al., 2008).

A commonly cited threshold or 'guard rail' for 'safe' levels of temperature rise is the so-called 2°C target as the upper limit of warming permissible to avoid dangerous anthropogenic interference in the climate (Randalls, 2010). The assumption is that a global mean temperature increase of up to 2°C relative to pre-industrial levels is likely to allow many human systems to adapt to climate change at globally acceptable economic, social and environmental costs (*EG Science*, 2008). The target has been advocated by the European Union as well as in the 2009 Copenhagen Accord that states: 'the increase in global temperature should be below 2 degrees Celsius'. However, the 2°C target dates back to the 1970s when scientists and economists sought to develop heuristics for policy decision-making on climate change (Randalls, 2010) – while the GHG emissions corresponding to a specified maximum warming are in fact poorly known (Meinshausen et al., 2009).

There are growing views that this threshold estimate might be too high (New et al., 2011) and not represent a reliable threshold. A 2°C target is committing the world to a significant degree of climate change, without much knowledge regarding whether or not it really prevents dangerous levels of climate change. Many negative effects of climate change are already observable under the current levels of global temperature increase (about 0.6°C to 0.7° over the period 1951–2010), and the adaptive capacity of many ecosystems might be well be exceeded *before* a 2°C increase is reached (EG Science, 2008). Attempts at estimating a GHG emission target for limiting global warming to 2°C suggest that limiting cumulative CO_2 emissions over the period from 2000–2050 to 1,000GtCO$_2$ (gigatons of CO_2) would yield a 25 per cent probability of global warming exceeding 2°C. Limiting cumulative CO_2 emissions to 1,440GtCO$_2$ over the same period would yield a 50 per cent probability of global warming exceeding 2°C (Meinshausen et al., 2009). This budget could easily be exhausted by 2027 or 2039 respectively (Meinshausen et al., 2009). However, the continued rise in GHGs along with a lack of a stringent and comprehensive global emission reduction agreement have made achieving a 2°C target extremely difficult (if not impossible) – which raises the likelihood of global temperature rises of 3°C or 4°C within this century (New et al., 2011). In our view, such a range would require significant and timely adaptive measures.

The Existence of Planetary Boundaries

Given the shortcomings of the 2°C target, new approaches emerged for estimating how dangerous levels of climate change and other types of global changes can be avoided. One such approach is the concept of 'planetary boundaries' (Rockström et al., 2009a; Rockström et al., 2009b). Rather than focusing on thresholds, the authors propose boundaries for human-induced pressures on the Earth's systems. Boundaries, in contrast to thresholds, are values of a variable (for example, temperature increases) set at a 'safe' distance from a dangerous level that should not be transgressed if unacceptable levels of global change are to be avoided. Determining this safe distance involves normative judgments of how much risk and uncertainty society is prepared to deal with. The authors identified nine planetary boundaries and offer quantifications for seven based on current scientific evidence. These seven boundaries are: climate change, ocean acidification, stratospheric ozone concentration, the biogeochemical nitrogen (N) and phosphorus (P) cycles, global freshwater use, land-use system change, and the rate at which biological diversity is lost. The proposition is that CO_2 concentration in the atmosphere should be lower than 350ppm, and/or that the maximum change in radiative forcing[3] should be $+1W/m^2$ (watts per square meter) – estimated by Hansen and Sato (2012) to result in a temperature change of approximately 0.75°C.

Looking at the existing evidence and the extent of already observable climate change, it is evident that the climate change boundary has already been transgressed. In addition to climate change, it is estimated that humanity has also transgressed boundaries for the rate of biodiversity loss and the rate of interference with the nitrogen cycle (driven by

---

[3] The IPCC (2007: 36) defines radiative forcing as follows: 'Radiative forcing is a measure of the influence a factor [e.g., the atmospheric concentrations of GHGs and aerosols, land cover, solar radiation] has in altering the balance of incoming and outgoing energy in the Earth-atmosphere system and is an index of the importance of the factor as a potential climate change mechanism. [...] radiative forcing values are for changes relative to preindustrial conditions defined at 1,750 and are expressed in Watts per square meter (W/m^2).'

agricultural activities and fertilizer use). Knowledge gaps and uncertainties exist in regards to the duration over which boundaries can be transgressed before causing unacceptable environmental change and the ability to return to a safe level once a dangerous level is reached (Rockström et al., 2009a; Rockström et al., 2009b). At the same time, however, approaching or transgressing any of the boundaries should not be seen as a fatalist excuse for inaction. First, decisive actions can reverse negative trends, as demonstrated with regard to the stratospheric ozone boundary. The signing of the Montreal Protocol (along with subsequent amendments) led to a reversal of a trend towards stratospheric ozone depletion (Rockström et al., 2009a). Decisive action on climate change could bring about similar successes. Second, interdependencies between boundaries need to be taken into consideration, suggesting that simultaneous actions should be undertaken. Third, the consequences of transgressing a boundary need to be evaluated, such that adaptive action can be undertaken to minimize impacts.

Scales of Change

Environmental problems are no longer localized. Scientific evidence increasingly points to global systematic changes in the environment (including climate change) that are linked to and regulated by a complex set of local processes (for example, local processes driving GHG emissions and land-use changes). At the same time, global changes have local and regional impacts and manifestations. The increasing extent of environmental problems beyond localized boundaries makes a consideration of both spatial and temporal scales important. 'Scale' in this context can be defined as a spatial, temporal, quantitative, or analytical dimension important for assessing problems and finding solutions (Cash et al., 2006; Gibson et al., 2000). A level within a scale refers to a unit of analysis located on a scale (often in a hierarchical manner). For instance, levels within a spatial scale could refer to patches, landscapes or regions up to the global level (Cash et al., 2006; Gibson et al., 2000).

The concept of scale is crucial for understanding interactions of organizations, industry and society with the natural environment. The first and most important concern is the increasingly global scale of adverse environmental consequences brought about by aggregate individual-level, organizational-level and industry-level activities. To date, however, the issue of scale in understanding the relation between organizations and the environment has been ignored or simplified (here defined in a broad sense to encompass both economic and environmental variables) (Gibson et al., 2000). Decision-makers within organizations

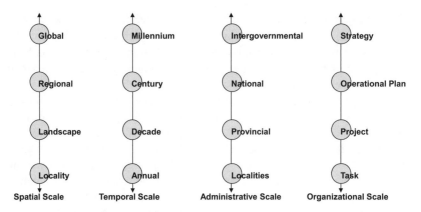

Source: Adapted from Cash et al., 2008.

Figure 1.2   Scales of change

tend to focus on relatively fast-moving variables on organizational and industrial levels (for example, financial resources, innovation capabilities, cost considerations, financial outcomes, or competitor moves), while the dynamics of a changing natural environment have typically not been considered in organizational decision-making. Business activities are often occurring in a way that is disconnected from the natural environment, not fully recognizing the importance of the natural environment, particularly its role in providing organizations with natural resources as operational inputs, a sink for waste, and stable, predictable natural environmental conditions (to the sectors that depend on this) (Linnenluecke and Griffiths, 2011). This book explores issues across a number of different scales and levels.

Administrative scales   Jurisdictional or administrative scales range from local administrative units (for example, local councils) to provinces or states, national governments and even intergovernmental levels. They are often closely related to spatial scales (see below) as they introduce a system of rules for certain geographic areas. In addition, jurisdictional or administrative scales often also introduce a system of rules between different levels, thus determining how rules or legislation are applied, introduced and amended over time (Cash et al., 2006). In the context of organizational responses to climate change, the importance of jurisdictional and administrative scales becomes evident when looking at institutional arrangements on state, regional and local level that introduce legislative or compliance requirements. Administrative requirements, especially if mandatory and enforced, are drivers governing organizational and industry responses to environmental issues (see also Chapters 3 and 4). In addition to mandatory requirements, some jurisdictions have voluntary programs targeting specific organizations or industries.

Beyond jurisdictional and administrative scales, businesses and other organizations have not been paying much attention to the issue of scale in decision-making, and especially in their response to climate change. For instance, organizations are typically not concerned with issues beyond their planning horizons. Organizations have also not been paying much attention to the relevance of geographic scale for economic activity beyond the importance of location for organizational growth over time. For example, few organizations have considered how the scale of economic activity relates to ecosystem dynamics (Brouwer, 2010; Linnenluecke and Griffiths, 2011).

Spatial scales   Spatial scales range from local to regional and global, and concern the geographic (spatial) extent of environmental and ecological phenomena, as well as organizational and industrial activities. Different levels on this scale and their interactions have been extensively studied in fields such as ecology (for instance, to understand ecosystems as well as their processes and changes at different levels). However, neither the geographic extent of organizational activities nor the concept of 'place' has received much attention in business research and practice. Business decisions on issues to do with geography or spatial scales are mainly related to economic considerations for the location of organizational activities (that is, which is the best location for production) and the expansion and 'scaling-up' of organizational activities (that is, organizational growth). Such considerations include the expansion into lucrative markets, increasing overall market size, responding to overseas competition, achieving cost-savings due to economies of scale, or gaining overseas cost advantages due to lower wages, resources access, or desirable infrastructure (Brouwer et al., 2004; Linnenluecke and Griffiths, 2011; Lloyd and Dicken, 1972).

The issues that have typically not been considered concern the impacts of the geographic scale and the scope of organizational activities on communities and the natural environment. Some organizations have been criticized for exploiting resources on a large scale to further their economic goals, paying little attention to the extent of resource use and resulting impacts on ecosystems, society and local communities. Such criticisms have been raised in particular about multinational corporations (MNCs), and especially by those opposed to globalization. Their concerns are that MNCs are stateless or rootless entities that exploit local places for the sole goal of maximizing shareholder value (Shrivastava and Kennelly, 2013). While mobility, global reach and presence in multiple locations have been theorized as essential elements of the competitive advantages of MNCs, these attributes also enable MNCs to tap into a variety of local (and, in particular, natural) resources and significantly scale-up not only their outputs, but also their overall environmental and carbon footprint (Patenaude, 2011; Shrivastava and Kennelly, 2013). Overall, the issue of scale within the business and management field has mostly been addressed in the context of scaling-up economic activities (that is, growth) and 'economies of scale' (that is, optimization of production to improve economic outcomes) (Gibson et al., 2000).

Temporal scales   Temporal scales consist of different 'time frames' and related durations, rates or frequencies. This book covers issues at different temporal scales, covering climate dynamics and change, policy developments, as well as organizational responses. Climate dynamics, for instance, are characterized by long durations and slower changes over decades to centuries, and have been excluded from many organizational decision-making frameworks and analyses altogether, especially those used in business decision-making (Egri and Pinfield, 1996; Shrivastava, 1994). However, even though climate dynamics take place over longer periods of time, they can manifest themselves in changes in the frequency and intensity of relatively short-lived phenomena such as weather extremes. Many institutional structures as well as governance structures or ideologies (worldviews) associated with current dominant economic paradigms are also characterized by 'longer' timeframes (Cash et al., 2006). Such slower-moving variables in the 'organizational environment' are consequently regarded as largely exogenous to the organization as they leave limited opportunities for individual organizations to influence them over shorter periods of time (Gibson et al., 2000). Timeframes most relevant for organizational decision-makers are short-term decision-making and goal-setting cycles at a strategic level (3–5 years). Operational targets can be even shorter and relate to daily, weekly, monthly or yearly outcomes.

Cross-scale and cross-level interactions   The real implications for organizational and industry responses to climate change arise not just from understanding the different scales and levels, but in particular from cross-scale and cross-level interactions and resulting challenges (Cash et al., 2006). This is important for understanding how global systematic changes and phenomena are linked to and regulated by a complex mix of more localized organizational and industrial processes (for example, business action on climate change as designed through strategies and carried out through different projects or tasks) and vice versa.

1. **Failure to recognize interactions between levels and scales**
    Cross-level and cross-scale interactions can be extremely difficult to understand. Examples include national policies that have unintended consequences on a local level or do not bring about desired changes (for instance, policies that constrain local action or innovation). Other examples are large-scale problems that arise due to the aggregate activities of individual actors (such as the rising concentration of GHGs due to emissions from a large number of distributed actors) or due to a focus on short-term outcomes (for example, profit optimization instead of long-term planning and sustainable solutions). As evident through the impacts brought about by the aggregate levels of industrial activities, local actions can give rise to significant levels of global change. In turn, however, industries and organizations will be impacted by local manifestations of these global changes.

2. **Mismatches between ecological systems and human (individual, organizational or industry) actions**   Many environmental problems are brought about by a mismatch between human actions (human scales) and ecological timeframes (ecological scales). Such mismatches can be brought about by the scale of resource use, where the aggregate level of use of a resource exceeds its local replenishment rate, or where short-term actions (for example, economic growth, resources consumption and electoral cycles) contradict long-term planning requirements. For instance, business and policy plans are often made in rather short-term decision-making cycles (3–5 years) that do not allow for long-term decision-making and adaptations. Furthermore, there is ample evidence that human environmental management attempts do often not match the scale of the problem (for example, transboundary environmental problems or migratory fisheries). Resulting challenges are matching the scale of knowledge on environmental problems and the scale at which decisions are made and action taken (Kates et al., 2001). This includes, for instance, translating scientific knowledge on global climate change processes into information that has relevance to local decision-makers, and using local, tacit or indigenous knowledge in national or international decisions.

3. **Difficulties to understand pluralities among actors, even at the same level**   Last, there is often the incorrect presumption that there is a single, correct, or best solution that should be applicable

to all actors at the same level. The simplest example is when a problem is defined as purely 'global' or 'local'. Climate change, for instance, is often framed exclusively as a global issue, which suggests that there needs to be a 'global' solution applicable on an international level. The impetus to frame issues at a single level comes from the need to both simplify and control. Governments, for example, often frame issues so that they can be managed within their jurisdictions. However, from an organizational perspective, finding a global solution is not seen as being of particular relevance. What matters are practical solutions that translate into competitiveness and innovation.

In the next chapter, we move from this preliminary analysis of the history of organizational and industry responses to the global environment to outlining the range of the threats posed by unabated climate change.

2.   The threat of climate change

The scientific consensus on climate change has been a much-questioned topic in public debate. Why then are the physical impacts of climate change becoming of such great importance for organizations and economies? Even though public opinion continues to be divided, assessments in the peer-reviewed scientific literature of the degree of consensus among publishing scientists on the topic of human-induced climate change point to strong consensus (Oreskes, 2004). In a recent study, Cook et al. (2013) analyzed the evolution of the scientific consensus on anthropogenic (human-induced) global warming in the peer-reviewed scientific literature. The authors examined a total of 11,944 abstracts of papers on the topics of 'global climate change' or 'global warming' in the period from 1991 to 2011. While 66.4 per cent of abstracts expressed no position regarding anthropogenic climate change (that is, did not address or mention the cause of global warming), 97.1 per cent of abstracts that expressed a position endorsed the consensus position that humans are causing global warming. Overall, the authors concluded that their analysis indicates the number of papers rejecting the consensus on climate change is a vanishingly small proportion of the published scientific research.

While the public often perceives that the scientific consensus on climate change is still debated, the fundamental scientific conclusion that humans are warming the climate is solid. Within the climate science field, some key uncertainties remain about the exact nature of climate impacts. One key uncertainty relates to how local conditions will change in addition to average global temperature rise. This uncertainty results from the models that are currently used to simulate the Earth's climate. These models are general circulation models (GCMs) that represent processes on a global level (that is, the global atmosphere, oceans, ice sheets and the land's surface). As these models focus on longer-term trends at larger scales, simply down-scaling the relevant data (in terms of time-scales and spatial scales) is a difficult exercise as errors might be introduced. These issues do not make regional simulations worthless, but their limitations and the implications for regional planning need to be understood (Schiermeier, 2010). These uncertainties do not take away from scientific consensus, but could mean that future changes could possibly be worse than currently projected (Schiermeier, 2010).

Intergovernmental Panel on Climate Change (IPCC)

One of the largest collections and assessments of worldwide scientific, technical and socio-economic information regarding climate change is undertaken by the Intergovernmental Panel on Climate Change (IPCC). The IPCC was jointly established by the United Nations Environment Programme (UNEP) and the World Meteorological Organization (WMO) in 1988, when climate change became a political issue. Its main purpose has been to assess the state of knowledge on climate change, including the science, the environmental and socio-economic impacts of climate change and also the possible response strategies. The IPCC issued assessments reports in 1990, 1996, 2001, and 2007, which had an important influence on the negotiators of the United Nations Framework Convention on Climate Change (UNFCCC) and the Kyoto Protocol (see Chapter 3). At the time of writing, the IPCC had launched the first and second part of its latest Fifth Assessment Report (AR5).[1]

The IPCC consists of three working groups (Working Group I–III) and a task force on national greenhouse gas (GHG) inventories. The working groups are responsible for collating scientific information on different issues related to climate change, ranging from the scientific aspects to adaptation and mitigation. Working Group I (WGI) is responsible for assessing the scientific aspects of climate change, and summarizes the scientific progress in understanding the human and natural drivers of climate change. The work of Working Group I is grouped around: *observation* (that is, observed changes in climate); *attribution* (climate processes and their causes); *projection* (estimates of projected future climate change); and *commitment* (the level of climate change that would continue even if GHGs were to be stabilized). Working Group I builds upon past IPCC assessments and incorporates new findings and scientific progress in each subsequent report (further information on WGI is available, accessed 20 September 2014 at https://www.ipcc-wg1.unibe.ch/). Working Group II investigates the vulnerability of human and natural systems as well as potential adaptation options (further information on WGII is available, accessed 20 September 2014 at

---

[1] New findings of the IPCC AR5 are included in Chapter 2 as far as they were available at the time of writing.

http://www.ipcc-wg2.gov/). Working Group III assesses options for limiting GHG emissions, mitigating climate change and related economic issues (further information on WGIII is available, accessed 20 September 2014 at http://www.ipcc-wg3.de/).

The key findings of Working Group I are that:

- 'Warming of the climate system is unequivocal, and since the 1950s, many of the observed changes are unprecedented over decades to millennia. The atmosphere and ocean have warmed, the amounts of snow and ice have diminished, sea level has risen, and the concentrations of greenhouse gases have increased' (IPCC, 2013). This finding is based on *observational evidence* of increases in global average air and ocean temperatures, rising global average sea level, and widespread melting of snow and ice.
- 'It is *extremely likely* (emphasis in original) that more than half of the observed increase in global average surface temperature from 1951 to 2010 was caused by the anthropogenic increase in greenhouse gas concentrations and other anthropogenic forcings together. The best estimate of the human induced contribution to warming is similar to the observed warming over this period' (IPCC, 2013). This finding is based on *attribution studies* evaluating whether and to what extent the observed changes occur due to human activity, or whether alternative physically plausible explanations exist.
- 'Continued emissions of greenhouse gases will cause further warming and changes in all components of the climate system. Limiting climate change will require substantial and sustained reductions of greenhouse gas emissions' (IPCC, 2013). This finding is based on studies that evaluate the *projected warming* under different emission scenarios.
- 'Most aspects of climate change will persist for many centuries even if emissions of CO_2 are stopped. This represents a substantial multi-century climate change commitment created by past, present and future emissions of CO_2' (IPCC, 2013). This finding is based on studies that attempt to assess the amount of warming that is 'committed' under the presumption that GHG emissions are stabilized.

The IPCC has estimated the likelihood of future climate change and trends in climate and weather extremes (see Table 2.1) based on emissions projections for the twenty-first century using emissions-based scenarios. Different future emissions projections were initially developed in the *Special Report on Emission Scenarios* (*SRES*), also referred to as the 'SRES scenarios' (Nakićenović et al., 2000). In the fifth and latest assessment report (*AR5*), the development of scenarios fundamentally changed from the IPCC-led *SRES* scenarios to the adoption of 'Representative Concentration Pathways', or RCPs (IPCC, 2013). Leading up to the publication of *AR5*, an *ad hoc* group of experts developed these RCPs as a more flexible, interactive and iterative alternative to the SRES scenarios (Moss et al., 2010; van Vuuren et al. 2011). The four RCPs correspond to a range of trajectories of GHG concentrations including different mitigation levels, whereas the *SRES* scenarios are policy independent. The RCPs are labeled by their approximate radiative forcing that is reached during or near the end of the twenty-first century (RCP2.6, RCP4.5, RCP6.0, RCP8.5).

Climate Change and Extreme Weather Events

Initially, much of the scientific debate focused on the gradual changes brought about by climate change and related impacts on ecosystems and society. The *IPCC First Assessment Report* (AR1), for instance, concluded that global mean temperature would be likely to increase by about 0.3°C per decade under a business-as-usual emissions scenario. These projections were changed as more knowledge on the Earth's climate became available and as climate change progressed. Projected likely increases in global mean temperature over the twenty-first century put forward in the 2007 IPCC report range between 1.1°C and 6.4°C (IPCC, 2007b). In addition, scientific consensus has established that climate change is already contributing, and will further contribute, to increases in the frequency and/or intensity of weather extremes (Meehl et al., 2007; Trenberth et al., 2007).

The projected trends, their estimated likelihood as well as examples of potential consequences for industries and organizations are summarized in Table 2.1. These projections show that organizations, industry and society (depending on their location) are likely to be exposed to changes in weather extremes, ranging from local and regional variations in weather patterns (for example, changes in agricultural yields due to more frequent hot days or heat waves) to large-scale and sustained changes across entire regions (for example, the inundation of some land areas due to extreme sea level rise). The potential combined occurrence of several extreme weather events, or several types of extremes across regions and sectors could be very damaging and further amplify the consequences of climate change.

Table 2.1   Future trends in weather extremes and projected impacts

Phenomenon and direction of trend	Assessment that changes occurred (since 1950s)	Likelihood of future trends	Examples of projected impacts on industries and organizations
Over most land areas warmer and fewer cold days and nights	Very likely	Virtually certain	Changes in agricultural yields, effects on water resources relying on snow melt, changes in energy demand, effects on winter tourism
Over most land areas warmer and more frequent hot days and nights	Very likely	Virtually certain	
Warm spells/heat waves frequency and/or duration increases over most land areas	Medium confidence on a global scale, but likely in parts of Europe, Asia and Australia	Very likely	Reduced agricultural yields in warmer regions, increased danger of wildfire, increase water demand and water quality problems
Heavy precipitation events. Increase in the frequency, intensity and/or amount of heavy precipitation	Likely more land areas with increases than decreases	Very likely over most of the mid-latitude land masses and over wet tropical regions	Damages to crops, soil erosions, inability to cultivate land, adverse effects on quality of surface and ground water, disruptions due to flooding, loss of property
Increases in intensity and/or duration of drought	Low confidence on a global scale, likely changes in some regions	Likely	Land degradation, lower yields, crop damage, increase livestock deaths, increased risk of wildfire, more widespread water shortages, possible migration away from vulnerable areas

Phenomenon and direction of trend	Assessment that changes occurred (since 1950s)	Likelihood of future trends	Examples of projected impacts on industries and organizations
Increases in intense tropical cyclone activity	Low confidence in long-term (centennial) changes, but virtually certain in the North Atlantic since 1970	Likely	Damage to crops, trees, coral reefs, power outages, disruptions caused by floods and high winds, withdrawal of insurance cover in certain areas, property losses, possible migration away from vulnerable areas
Increased incidence and/or magnitude of extreme high sea levels	Likely (since 1970)	Likely	Impacts on water supplies, costs for coastal protection and relocation, possible migration away from vulnerable areas

Notes: 'Virtually certain' >99 per cent probability of occurrence; 'Very likely' >90 per cent probability; 'Likely' >66 per cent probability.

Sources: Adapted from IPCC data (IPCC, 2013; IPPC, 2007).

Based on statistical reasoning, changes in extremes can result from relatively small changes in the mean, variance, or shape of probability distributions, or all of these factors. Extremes occur at the high and low end of the range of values of a particular variable. For example, temperature extremes occur at the high and low end of the temperature range. The probability of occurrence of values in this range is called a *probability distribution function*. For some variables, such as temperature, this function is shaped similarly to the so-called bell curve. Figure 2.1 illustrates the probability distribution of temperature and illustrates the effect a small shift (corresponding to a small change due to global warming) can have on the frequency of extremes at either end of the distribution. Changes in the variability or shape of the distribution are not reflected in this simple picture. Some extremes, such as droughts, may be the result of an accumulation of different variables that might not necessarily be extreme when considered independently (Solomon et al., 2007a).

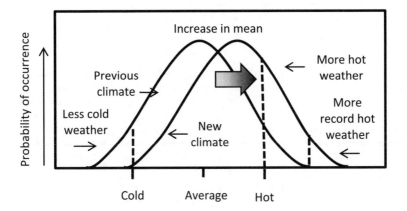

Notes: Illustration showing the effect of mean temperature increase on extreme temperatures (assuming a normal temperature distribution)

Source: Adapted from Solomon et al. (2007a).

Figure 2.1   The effect on extreme temperatures when the mean temperature increases

While many extreme weather and climate events continue to be the result of natural climate variability (IPCC, 2012), observational evidence suggests that the likelihood of some extreme events, such as heat waves, has already increased as a result of climate change, while the likelihood of other extremes, such as extremely cold nights, has decreased (Hegerl et al., 2007). Scientists have undertaken studies to empirically determine the contribution of human activities to extreme events, such as the 2003 European heat wave and hurricanes Katrina and Rita (Stott et al., 2004; Trenberth, 2005). These studies do not seek to establish a connection between climate change and weather extremes in a simple deterministic sense, as human-induced influences on the Earth's climate, such as elevated concentrations of GHGs in the atmosphere, are not simply causing an extreme weather event. Existing studies (summarized in Table 2.2) have produced findings that human activities may have already increased the risk of occurrence and/or intensity of some extremes. In other words, while the occurrence of any single past weather extreme cannot be directly linked to climate change alone, scientific research findings strongly suggest that human influences have a contributing role.

Table 2.2   Human contributions to extreme weather events

Extreme weather event	Research findings (examples)
2003 European heat wave	• It is very likely (confidence level >90 per cent) that human influence has at least doubled the risk of European mean summer temperatures as hot as in 2003 (Stott et al., 2004).
2011 flooding in the UK in autumn 2000	• It is very likely that global anthropogenic GHG emissions substantially increased the risk of flood occurrence in England and Wales in autumn 2000 (Pall et al., 2011).
Hurricanes	• Trends in human-influenced environmental changes are evident in hurricane regions and expected to affect hurricane intensity and rainfall. The effect on hurricane numbers remains unclear (Trenberth, 2005; Trenberth and Shea, 2006).
Australia's 'Angry Summer' of 2012–13	• Extreme heat waves and catastrophic bushfire conditions during the Australian summer of 2012–13 were made worse by climate change (Steffen, 2013).

Evidence is now showing that a 'new normal' for many climate- and weather-based (extreme) indices has been reached (Geneva Association, 2013; IPCC, 2012). Consequently, research is increasingly focusing on researching the impacts of climate scenarios beyond a 2°C warming outcome. The IPCC has established that there will be large variations in spatial patterns of climate change associated with any global temperature increase. For instance, land areas warm more than the oceans. Consequently, temperature increases will exceed the global average (frequently by more than one-and-a-half times) in areas inhabited by humans. And while average global precipitation is projected to increase, variations can also be expected in regards to rainfall patterns. Most areas that are currently arid and semi-arid are projected to dry, while the mid-latitudes and moist tropics and are projected to become wetter. Such trends have already been detected in recent observations of precipitation (New et al., 2011).

Disasters in a Changing Climate

Weather extremes resulting in a large number of people killed or injured, or in major economic losses, are commonly referred to as climate- or weather-related disasters (Brooks and Adger, 2003). The IPCC Report *Managing the Risks of Extreme Events and Disasters to Advance Climate Change Adaptation* (IPCC, 2012) considers the implications of extreme weather and climate events (climate extremes) for society (and sustainable development in general) in further detail. The report sees the implications of future changes in extreme weather events as particularly impactful if they coincide with increased exposure (that is, the presence of people or infrastructure in places that could be adversely affected) and vulnerability to climate change (that is, the propensity or predisposition of people or infrastructure to be adversely affected). The report cites a number of factors that will further contribute to climate- or weather-related disasters and losses:

- extreme weather events will be more impactful on sectors which are more closely linked to climate, including water, agriculture, health and tourism;
- in many regions, the main drivers of economic losses will be socioeconomic changes due to changes in population and increased exposure and vulnerability of people and assets; and
- disasters associated with climate extremes will influence population mobility and lead to relocation affecting both host and origin communities.

The report makes a case for adaptation to climate change, defined as adjustment to actual or expected climate and its effects, with the aim to moderate harm through a reduction of exposure and vulnerability, or to even exploit beneficial opportunities. At the same time, the report recognizes that not all disasters can be avoided through adaptation alone. Consequently, the report also makes a case for strengthening the resilience of human systems (such as settlements, industrial systems), defined as 'the ability of a system to anticipate, absorb, accommodate, or recover from the effects of a hazardous event in a timely and efficient manner, including through ensuring the preservation, restoration, or improvement of its essential basic structures and functions' (IPCC, 2012: 5).

Implications for Organizations

Climate-related disasters occurring under current climate conditions already illustrate the often significant consequences of such disasters for affected organizations and industries, and the social and/or economic stability of an entire region. These events can serve as analogues for future events (and are likely, or very likely, to change in frequency and/or intensity by the end of the twenty-first century) (Linnenluecke and Griffiths, 2012). The 2003 European heat wave, with anomalous hot and dry conditions, is often cited as an analogue for future climate change. The heat wave led to record drops in agricultural yield (up to 36 per cent in countries such as Italy) and had considerable impacts on various sectors ranging from health services, water supplies, food storage, and energy production. In France, six power plants had to be shut down due to increased energy demand and diminished cooling efficiency (Wilbanks et al., 2007b).

One Degree Matters

To visualize the impacts of different degrees of temperature rise to decision-makers, the European Environmental Agency produced a film titled *One Degree Matters*. The film highlights that a global average temperature increase of 2°C is expected to lead to coral reefs dying back, pressures on animals and plants to migrate or face extinction, negative impacts on livestock and crops (due to more frequents droughts and floods), the failing of some ecosystems and overall less security for societal infrastructure. A global average temperature increase of 3°C is expected to lead to greater flooding, more intense storms, changes to the patterns of monsoons and longer periods of drought. It is anticipated that low-lying areas around the world will be regularly flooded due to sea level rise and storm surges, unless extensive coastal defenses have been erected. A warming of that extent is also expected to have significant impacts on the melting of sea ice. In the Arctic, sea ice is projected to virtually disappear, and methane (a GHG that is 20 times more powerful than CO_2) will be lost at an accelerating rate from thawing permafrost, thus further contributing to global warming. Glaciers that are important for the provision of water resources are projected to disappear as well. A likely outcome of 3°C warming is an irreversible melting of large sections of the Greenland Ice Sheet.

The expected changes highlighted for a warming of 4°C include the instability of ice sheets in Antarctica, widespread species extinction, further rises in sea levels, large-scale dying of coral reefs due to ocean acidification and warming. Furthermore, it is anticipated that there will be soil losses and aquifer depletion, meaning that agriculture is no longer reliable, which will cause significant pressures to feed a growing population. Possible outcomes are large-scale population migration to areas with greater food security. The last scenario presented by the film is a global average temperature increase of 5°C warming which is expected to lead to unrecognizable changes, including the disappearance of major ice sheets and areas of rainforest, the inundation of coastal cities, as well as rising inland temperatures of 10°C or more. This could lead to consequences such as mega-droughts as well as large-scale population migration into areas less affected by pressures from areas of intense droughts and floods.

Important implications will arise from differences in pattern of change between a 2°C and 4°C world. Given that climate models vary widely in their projections of both global mean temperature rise and regional outcomes, studies have explored if there are any systematic differences in regional changes associated with 'high-end' warming outcomes, that is, models projecting 4°C or more by the end of the twenty-first century relative to the pre-industrial era (Sanderson et al., 2011). While the pattern of warming relative to global mean temperature change is similar between high-end models and those with less projected warming, differences arise in terms of the amplification of local changes. For instance, under a high-end warming outcome, northern Africa is projected to experience high (greater than 6°C) temperature increases and large precipitation decreases, suggesting that this region is among those most at risk.

One of the most certain outcomes associated with a greater average global temperature is the amount of projected sea level rise (New et al., 2011). The range of sea level rise for a temperature rise of 4°C or more over the same time frame is projected to be between 0.5 m and 2 m (Nicholls et al., 2011). However, uncertainties exist, in particular concerning the irreversible melting of parts of the Greenland ice sheet and a possible break-up of the West Antarctic ice sheet. Other thresholds might be crossed as well under a higher level of global warming as portrayed in the film documentary *One Degree Matters*.

The need for organizational transformation
While projections of a radically changing environment might seem like 'doom and gloom' scenarios to an organizational decision-maker, they are demonstrating the very real risks of a climate-changed world, especially if climate change remains unmitigated and large-scale impacts on ecosystems, economies, and society occur as a result. Such large-scale changes will require a shift from incremental adaptation in organizational practices to the need for transformative change and the uptake of resilience measures. Large-scale consequences can be anticipated in supply chains, including changed access to raw materials, changed timing and location of production as well as changed access to distribution channels, customers and markets (Linnenluecke and Griffiths, 2011). Consequences may also result from stakeholder and policy pressures. Organizations will require new strategies that address a wide variety of these issues as they face real and destructive impacts.

Dependency on climate-sensitive or non-renewable natural resources Businesses or markets are constrained by, and dependent upon, ecosystems and natural resources. As the impacts of climate change increase, organizations that are highly dependent on climate-sensitive or non-renewable natural resources will require strategies that take this dependency into consideration (Hart, 1995, 1997). Possible options include lowering the dependency on climate-sensitive or non-renewable natural resources (substitution, efficiencies), lowering the dependency on external suppliers (vertical integration) or responding to possible disruptions in supply (through flexibility and redundancy, such as diversification across supply sources).

Vulnerability to external changes   In 2013 the World Economic Forum has ranked failure of climate change adaptation among the top five global risks (World Economic Forum, 2013). If unmitigated, the expected consequences of climate change could lead to widespread impacts on organizations across a range of industry sectors and lead to significant economic losses, but also rising food insecurity, aggravated water scarcity and population displacements. Beyond individual organizations, supply chains are very vulnerable to increases in the number and/or severity of weather-related disruptions, especially to resources inputs and the movement of products and resources throughout a country and internationally.

Emerging climate legal risks   Specific climate change adaptation laws are only just emerging across countries and globally. Climate legal risk, that is, legal risk that arises from changes in the natural environment, is not an easily defined term. It spans a range of possible legal consequences for organizations, including the possibility of changes to corporate law, changed regulatory and reputational risks, insurance risks and changes to existing legal frameworks, such as changes to planning schemes or zoning laws (Johnston et al., 2013). Several court cases have already raised the issue of climate change, for instance over development consent for proposals for coal-fired power plants or coal mines, as well as the extent to which climate change impacts (projected sea-level rise, coastal erosion, or flooding) need to be considered in coastal planning processes (Preston, 2011).

To date, claimants face substantial challenges. Not only do they need solid scientific evidence, but the non-scientific principle has not yet been established in regards to what kind of evidence on climate impacts is admissible and how liability must be established. Claims for actual damage associated with climate change are likely to involve the kind of evidence on the human contributions to extreme weather events presented above as they seek to establish how weather risks would have been different if human influence had not occurred. If courts accept such evidence, the role of human influence in some extreme weather events is already likely to exceed the threshold at which some form of liability might apply (Allen, 2010). Questions are likely to arise in terms of attribution of liability – but also in regards to tortious liability for government entities, developers and landholders (Preston, 2011).

Investment strategies Climate change impacts are also gaining import-ance in investment decisions by large asset owners (pension and super-annuation funds, foundations, endowments, insurance companies and wealth funds), in particular due to interest in quantifying the strategic implications of climate policy as well as physical risks to long-term investments. Several investor groups have formed around understanding the nature of risks and opportunities, including the Global Investor Coalition on Climate Change, which consists of four regional investor groups: the Institutional Investors Group on Climate Change (IIGCC) in Europe; the Investor Network on Climate Risk (INCR) in North America; the Investor Group on Climate Change (IGCC) in Australia and New Zealand; and the Asia Investor Group on Climate Change (AIGCC) in Asia). Other investor groups include the Carbon Disclosure Project (CDP) or the Asset Owners Disclosure Project (AODP). The AODP estimates that the 1000 largest assets owners manage over US$60 trillion of funds. While their recent survey results suggest that no single asset owner has accurately assessed or managed climate risk over the entire portfolio, leading organizations are making a start by monitoring their climate and carbon exposure (for further information, see aodproject.net, accessed 21 September 2014).

Insurability Due to the anticipated increase in the frequency and/or intensity of extreme weather events, climate change will have adverse impacts on insurance affordability and availability. This, in turn, could shift the burden of having to cover for losses to governments and individuals (Mills, 2005). Issues around insurability are already material-izing. For instance, in Australia, many insurers have significantly increased insurance premiums after recent disastrous weather extremes. The ramifications of insurance availability may ultimately filter through to the lending and mortgage sectors. Mortgage viability may become an issue in locations that are repeatedly impacted by extreme weather events and have experienced significant urban growth and expansion into vulnerable locations (Johnston et al., 2013).

Stakeholder pressure Organizations will also face increased external pressures from stakeholders demanding action on climate change. In a survey on adaptation of public- and private-sector organizations (conducted by Gardner et al., 2010), five of the 242 participating organizations reported that they felt pressure from governments and local communities to adapt to climate change. This pressure included the perceived need to comply with specific government policies, newer building and engineering codes, as well as demands by community members and landholders. One organization in the survey felt that it would be ineligible for inclusion in certain investment funds unless it addressed climate change adaptation. While five out of 242 organizations might seem like a small percentage, it can be expected that such pressures will increase as the impacts of climate change become more visible.

Systemic risks A last critical point is a heightened level of systemic risk due to climate change impacts on critical infrastructures. Sectors such as water, energy, and telecommunications are critical for the functioning of modern society. However, these complex and interconnected critical infrastructure (CI) systems are widely considered to harbor an increased propensity for breakdowns. While such breakdowns have been rather rare overall, the impacts of hurricanes Katrina and Rita on the Gulf of Mexico, or Hurricane Sandy on the Caribbean and East Coast of the US have demonstrated the catastrophic consequences of such breakdowns, and have shown how vulnerable modern society is to impacts from the natural environment. After the impact of hurricanes Katrina and Rita, energy company Entergy New Orleans experienced significant supply disruptions. The company required 45 days to restore power to its customers. This operation proved to be extremely costly since a vast amount of infrastructure had to be repaired and rebuilt (Entergy, 2005; Linnenluecke et al., 2012). Since then, Entergy has begun to incorporate the risks from climate change and weather extremes into its business planning and operating activities, and has started to strengthen highly vulnerable sites within its power distribution network (Coffee, 2014). Such examples show that there is a heightened sense of vulnerability to climate change as a new and future threat (Boin and McConnell, 2007).

Under a rapidly transformed environment, especially those organizations that are highly dependent on natural resources might find themselves exposed to the severe impacts of climate change on their infrastructure (agriculture, transportation), discover limitations in their business models (insurance, reinsurance and financial services), or might experience consumer boycotts of their products and services if their practices are found to be damaging the fragile ecosystems of the Earth and creating negative social impacts for communities. Responses to these changes will require large-scale and transformational action with changes to existing business models, structures and practices. As the impacts of climate change become more visible it can also be expected that the regulatory environment targeting GHG emissions will become more stringent.

CASE STUDY: CLIMATE CHANGE AND THE INSURANCE INDUSTRY

The Travelers Companies, Inc. is a provider of personal and commercial property insurance products in the US. Following the Atlantic hurricanes in 2004 and 2005, the company developed an integrated approach to climate change risks by forming several internal working groups on climate change (Sussman and Freed, 2008).

Travelers has taken the following actions within their existing business models:

- Reassessment of coastal underwriting practices by engaging in loss control, and adaptation activities within coastal locations.
- Updating catastrophe modeling to ensure better integration of climate change science within these models.
- Offering risk control services by monitoring building code standards and regulations to build resiliency, and providing assistance in disaster-preparedness planning.
- Redesigning pricing to take into consideration differences in factors such as building age, loss mitigation efforts, and other factors that affect likely losses during extreme weather events.
- Engaging in community and government outreach by providing information to governmental departments about the benefits of long-term loss mitigation strategies such as the adoption of more robust building codes (Sussman and Freed, 2008).

Travelers has also worked with other organizations to implement adaptation measures. For example, Travelers provides financial support to the Institute for Business and Home Safety in order to support their building code enforcement training programs for local, and state building inspectors, and private contractors in Louisiana (Sussman and Freed, 2008).

This chapter began by asking why the physical impacts of climate change are becoming of such great importance for organizations and economies. In large part, the answer is that the new reality of climate change brings about significant risks, linking adaptation to climate change with organizational success. It is likely that some organizations will be reacting primarily to the reputational and litigious risks associated with climate change. Others may take more proactive measures, seeking to address both adaptation and mitigation in an effort to conserve resources, and limit adverse impacts. Importantly, investors will be looking to more than financial success in evaluating the future profitability of investments. The selection of investments will increasingly take into account longer-term factors such as possible adverse increases in the capacity impacts of a changed environment onto assets. In this context, organizations require an integrated approach to manage climate risks with a view to reduce their vulnerability and maintain or increase the opportunities for sustainable development. One of the ongoing difficulties that organizational decision-makers face in responding to issues of global impacts is the time of their tenure. Most executives have a limited time running businesses, and their decision-making is thus characterized by short-term strategies.

The following chapters outline the external policy and institutional context for corporate responses on climate change, along with the possible spectrum of corporate responses. We begin by canvassing the international climate policy landscape.

3. The international climate policy landscape

Climate change was long not considered as an issue of international political importance. It was only in 1992 with the signing of the United Nations Framework Convention on Climate Change (UNFCCC) that international political efforts, and national climate policies slowly started to emerge. This chapter provides a short timeline of climate policy negotiations, focusing on key policy developments on international levels. Implications for policies on a national level are discussed in subsequent chapters. It has to be noted that importance of developments such as the UNFCCC, and the Kyoto Protocol is not in just their direct, but also in their indirect (flow-on) effects on national, industry, and organizational-level action on climate change.

A closer look at the history of adaptation within the UNFCCC process shows that the original main intention of the UNFCCC treaty was to focus on mitigation efforts, and the development of emission reduction targets to 'prevent dangerous anthropogenic interference with the climate system' (United Nations, 1992). International negotiations mainly centered on country-level targets for greenhouse gas (GHG) emission reductions or mitigation measures, leaving it by and large up to individual countries to design policy instruments and mechanisms to target specific industry-sectors or high-emitting activities. It was only since the early 2000s that the focus of climate policy negotiations broadened to consider adaptation to the impacts of climate change.

The Convention text does include some references to the need to prepare for and fund adaptation, especially in developing countries particularly vulnerable to the adverse effects of climate change. However, adaptive capacity was initially seen as the extent to which society could adjust to, and could tolerate change in climate, and was not defined as a policy objective *per se*. However, the second, post-2000 wave of policy negotiations reflected a greater attention to adaptation as a response strategy to climate change, alongside mitigation. As a result, and as concerns about adaptation have only just begun to emerge as a more prominent topic on the international climate policy landscape, adaptation objectives have not yet been integrated into key policy objectives, leading to uncertainties in the way forward. By outlining the political background to action on both mitigation of and adaptation to climate change, this chapter is setting the scene for subsequent chapters on national, organizational, and industry level responses to climate change.

International Policy Developments

Even though scientists started to pay greater and more systematic attention to climate change, and global warming from the 1950s onwards, these issues largely remained topics within scientific debates. It was not until the 1990s that scientific concerns about global warming attracted policy attention (see Table 3.1). In 1979, scientists who organized the First World Climate Conference attempted to attract participation by policy decision-makers, but efforts proved largely unsuccessful. Subsequent international scientific and political meetings such as the 1988 Toronto Conference, the 1989 Hague and Noordwijk conferences and the 1990 Second World Climate Conference slowly attracted some attention from ministers and heads of government. Climate change was eventually raised as an issue in the UN General Assembly (Bodansky, 2001; Gupta, 2010). In 1988, the World Meteorological Organization (WMO) and the United Nations Environment Program (UNEP) established the Intergovernmental Panel on Climate Change (IPCC) in an attempt to bring together international climate experts to synthesize the most recent climate science findings and assess the state of global knowledge on climate change.

The United Nations Framework Convention on Climate Change (UNFCCC)

The IPCC First Assessment Report in 1990, together with the outcomes of the Second World Climate Conference held in the same year, strongly supported policy action on climate change. These developments led to intergovernmental negotiations, and the establishment of the Intergovernmental Negotiating Committee for a Framework Convention on Climate Change (INC/FCCC) by the UN General Assembly. The INC/FCCC eventually adopted the United Nations Framework Convention on Climate Change (UNFCCC) at the United Nations Conference on Environment and Development (also referred to as 'The Earth Summit') held in Rio de Janeiro, Brazil, in 1992 (United Nations, 1992).

Table 3.1   Summary overview of international climate negotiations

	Pre-1990	1990–1996	1997–2001	2002–2007	2008–2012	2013 and beyond
Key focus	Climate change as environmental issue		Climate change as developmental issue		Climate change as adjustment issue for economies and society	
Scientific basis	First and Second World Climate Conference	IPCC First Assessment Report	IPCC Second Assessment Report	IPCC Third Assessment Report	IPCC (2007) Fourth Assessment Report; SREX Report (IPCC) (2012)	IPCC Fifth Assessment Report
Key policy developments		UNFCCC (1992) (entered into force in 1994); Berlin Mandate (1995)	Kyoto Protocol (1997); Marrakech Accords (2001)	Start of European Emissions Trading Scheme (EU ETS) (2005); Bali Roadmap (2007)	First Commitment Period under Kyoto Copenhagen Accord (2009); Cancun Agreements (2010); Durban Platform (2011); Doha Amendment (2012)	Second Commitment Period under Kyoto; Adoption of a universal climate agreement by 2015 *(to come into effect by 2020)?*; Warsaw (2013)
Scope		Three gases (carbon dioxide, methane, nitrous oxide)	Kyoto: Six gases (carbon dioxide (CO_2), methane (CH_4), nitrous oxide (NO_2), hydrofluorocarbons (HFC), sulphur hexafluoride (SF_6) and perfluorocarbons (PFC)), sinks, offset mechanisms			Six Kyoto gases and nitrogen trifluoride (NF_3)
			Bali Roadmap introduces NAMAs (Nationally Appropriate Mitigation Actions) for developing countries			

	Pre-1990	*1990–1996*	*1997–2001*	*2002–2007*	*2008–2012*	*2013 and beyond*
Countries captured by Agreements		UNFCCC: Division into developed countries (OECD incl. JUSS-CANNZ and EC-12/EC-15, plus EIT) and the developing world (G77 and China, AOSIS, OPEC)	Kyoto: Division into Annex B Parties (Annex I Parties minus Turkey, Belarus) and non-Annex B Parties	Australia ratifies Kyoto Protocol	Annex B countries with binding targets during first commitment period	EU, Australia and other developed countries with binding targets during second commitment period
Influential negotiating coalitions	European Communities and AOSIS (advocating CO_2 reductions); US and OPEC (opposed); Brazil, India, and China (concerned about sovereignty)	OECD incl. JUSS-CANNZ and EC-12/EC-15, plus EIT countries G77 and China, AOSIS, OPEC	EC-15, Umbrella group (former JUSS-CANNZ) G77 and China, AOSIS, OPEC	EU, Umbrella Group, Central Group-11 (most EITs included in Annex I), EIG; G77 and China, AOSIS, OPEC, GRULAC, African Group, CACAM	In addition to existing coalitions: US, and BASIC (Copenhagen) Cartagena Dialogue (Cancun)	Visions of a global climate deal; Developing countries gaining in importance, rifts showing in G77/China
Long-term climate target at time		Prevention of 'dangerous anthropogenic interference' with the climate system	Annex B Parties to jointly reduce total emissions of six GHGs by 5.2% (period 2008–12)			Successor to Kyoto
					2°C target, pressure from developing countries for 1.5°C target	

	Pre-1990	*1990– 1996*	*1997– 2001*	*2002– 2007*	*2008– 2012*	*2013 and beyond*
Key policy focus		Mitigation	Mitigation, development of Kyoto mechanisms (ET, CDM, JI)	Mitigation, launch of Adaptation Fund in 2007 (first projects approved in 2012)	Mitigation (ET, CDM, JI), adaptation: launch of UN REDD in developing countries	Second- and third-generation biofuels, geo-engineering, human rights, adaptation, REDD, CDM/ET

Sources: *Adapted from Bodansky (2001) and Gupta (2010).*

The UNFCCC entered into force in 1994 as an international treaty to allow countries to cooperatively consider what they could do to limit climate change, and to minimize the progression of impacts that were seen as already inevitable. The UNFCCC was set up with the ultimate goal to stabilize GHG concentrations 'at a level that would prevent dangerous anthropogenic (human induced) interference with the climate system'. It states that: 'such a level should be achieved within a time-frame sufficient to allow ecosystems to adapt naturally to climate change, to ensure that food production is not threatened, and to enable economic development to proceed in a sustainable manner'.[1] Under the Convention, participating governments agreed to collect and share information on GHG emissions (national GHG emissions inventories), national policies and best practice approaches; to launch national strategies for addressing GHG emissions and adaptation, including the provision of financial and technological support to developing countries; and to cooperate in preparing for climate change adaptation.

The UNFCCC initially classified countries into developed countries (Annex I Parties) and developing countries (non-Annex I Parties) following political tradition at the time. A key guiding principle was introduced in Article 3 of the 1992 UNFCCC document:

The Parties should protect the climate system for the benefit of present and future generations of humankind, on the basis of equity and in accordance with their *common but differentiated responsibilities and respective capabilities*. Accordingly, the *developed country Parties should take the lead in*

---

[1]   For a critique of the objectives, see Chapter 1.

combating climate change and the adverse effects thereof. (United Nations, 1992; emphasis added)

Included in Annex I were those industrialized countries that belonged to the Organization for Economic Co-operation and Development (OECD) in 1992, plus 12 countries with 'economies in transition' (the EIT Parties) from Central and Eastern Europe, including the Russian Federation, the Baltic States and several Central and Eastern European States (UNFCCC, 2007b). Annex I Parties were expected to engage in greater emission cuts through national policies and mitigation strategies compared to non-Annex I Parties – as they were seen as being responsible for most of the increase in human-induced GHG emissions, and were expected to reduce emissions by the year 2000 to 1990 levels (UNFCCC, 2013j). The Annex I Parties also agreed to develop, publish, and regularly update national GHG emission inventories. However, as the economies in transition faced significant financial difficulties, a new Annex (Annex II) of richer developed countries was established (Gupta, 2010).

Included in Annex II were the OECD member countries among the Annex I Parties, but not the EIT Parties. Annex II Parties were expected to provide funding to support developing countries in undertaking emission reduction activities under the Convention. In addition, Annex II Parties were also expected to assist developing countries in their efforts to adapt to the adverse effects of climate change. Furthermore, Annex II Parties agreed to 'take all practicable steps' to promote the development and transfer of environmentally-friendly technologies to developing countries as well as EIT Parties (UNFCCC, 2013j). Non-Annex I Parties consisted mostly of developing countries. Some developing countries were classified by the Convention as being particularly vulnerable to the impacts of climate change, such as countries with low-lying coastal areas, and those already affected by drought and desertification. The Convention also recognized that other countries (such as those heavily reliant on income from fossil fuel production) might feel more vulnerable to the potential economic implications of measures to respond to climate change (UNFCCC, 2013j). The UNFCCC classified 49 Parties as least developed countries. These countries were given special consideration under the Convention due to their limited capacity to respond and adapt to climate change impacts (UNFCCC, 2013j).

The UNFCCC did not commit member countries to mandatory GHG emission reductions and therefore contained no enforcement require-ments. All developed countries (except Turkey) thus ratified the UNFCCC quickly, in particular because it did not impose any binding obligations, and it entered into force in 1994. The Convention's central body is the Conference of the Parties (COP), which comprises delegates of member states that have ratified or acceded to the agreement. The first annual meeting of Parties, or the First Conference of the Parties (COP-1), took place in Berlin, Germany. The first negotiated outcome, the 'Berlin Mandate', addressed that the UNFCCC lacked a binding agreement on emission reductions. The mandate interpreted the principle of 'common but differentiated responsibilities and respective capabilities' in Article 3 of the UNFCCC to mean different commitments for Annex I and non-Annex I Parties. The decisions of the COP-1 included to:

- 'begin a process to enable it to take appropriate action for the period beyond 2000, including the strengthening of the commitments of the Parties included in Annex I to the Convention (Annex I Parties) … through the adoption of a protocol or another legal instrument';
- 'elaborate policies and measures [for developed country/other Parties included in Annex I], as well as to set quantified limitation and reduction objectives within specified time-frames, such as 2005, 2010 and 2020, for their anthropogenic emissions by sources and removals by sinks of greenhouse gases'; and 'not introduce any new commitments for Parties not included in Annex I'

One of the significant outcomes of COP-1 was the decision to set up an *ad-hoc* committee for negotiating a protocol or other legal instrument to be adopted at the Third Conference of the Parties (COP-3) in 1997 at Kyoto with binding commitments by industrial countries to reduce their GHGs after the year 2000. This led to more than 30 months of international negotiations, the so-called 'Kyoto Process' (Oberthür and Ott, 1999). The Kyoto Process faced strong opposition from its outset, in particular also from the US Senate. The US Senate responded to the Berlin Mandate by unanimously passing the (not legally binding) Byrd-Hagel Resolution in 1997 (US Senate, 1997) which stated:

> It is the sense of the Senate that the United States should not be a signatory to any protocol to, or other agreement regarding, the United Nations Framework Convention on Climate Change of 1992, at negotiations in Kyoto in December 1997, or thereafter, which would mandate new commitments to limit or reduce greenhouse gas emissions for the Annex I Parties, unless the protocol or other agreement also mandates new specific scheduled commitments to limit or reduce greenhouse gas emissions for Developing Country Parties within the same compliance period.

While progress towards deciding on binding emission reduction targets was slow initially, the adoption of the Geneva Ministerial Declaration in July 1996 at COP-2 reinforced the outcomes of COP-1 and endorsed the significantly strengthened Second Assessment Report (AR2). The AR2 concluded that: 'The balance of evidence suggests a discernible human influence on global climate', and included additional materials on the implications of various potential emission limitations and regional consequences. Importantly, COP-2 did not attempt to have the Declaration adopted by majority vote, thus showing the willingness of countries to act in the absence of international consensus. COP-2 thus merely took note of the Declaration and appended it to the final report, over the opposition of Saudi Arabia (and other OPEC states), Russia, and Australia (Bodansky, 2001; Gupta, 2010).

The Kyoto Protocol to the UNFCCC

The negotiations following the Berlin Mandate provided the basis for the Kyoto Protocol to the UNFCCC. The Kyoto Protocol was adopted in 1997 at COP-3 as an extension to the Convention that outlined binding commitments to emissions. While the UNFCCC only encouraged countries to reduce their emissions, the Kyoto Protocol committed industrialized countries to stabilize GHG emissions based on the principles of the Convention. The Kyoto Protocol covers six major GHGs: carbon dioxide (CO_2), methane (CH_4), nitrous oxide (NO_2), hydrofluorocarbons (HFCs), perfluorocarbons (PFCs) and sulphur hexafluoride (SF_6). The Kyoto Protocol set binding emission reduction targets for 37 industrialized countries and the European Union in its first commitment period based on the principle of 'common but differentiated responsibility'. The dichotomous Annex I/Non–Annex I distinction introduced by the UNFCCC thus has had significant implications on international mitigation targets (Aldy and Stavins, 2012). Overall, the targets under Kyoto add up to an average 5 per cent in emission reductions compared to 1990 levels over the five-year period from 2008 to 2012 (the first commitment period) (UNFCCC, 2013h).

The next round of negotiations at COP-4 in Buenos Aires, Argentina, concerned implementation issues, including finance and technology transfer. In 2001, governments reached a political deal – the Bonn Agreements – signing off on the controversial aspects of the Buenos Aires Plan of Action. The release of the *IPCC Third Assessment Report* supported negotiations by offering the most compelling scientific evidence on global warming thus far (UNFCCC, 2007b).

The detailed rules for the implementation of the Kyoto Protocol were adopted at COP-7 in Marrakech, Morocco, in 2001 (also referred to as the 'Marrakech Accords'). The Kyoto Protocol eventually entered into force on 16 February 2005 due to a complex ratification process (UNFCCC, 2013h). Both the US and Australia did not ratify the Kyoto Protocol at the time; the US, responsible for about 25 per cent of the 1990 carbon dioxide emissions, withdrew in 2001 after the election of George Bush.[2] Russia ratified the Kyoto Protocol in 2004, satisfying the condition that it only enters into force when signatories include industrialized countries responsible for 55 per cent of the developed world's carbon dioxide (CO_2) emissions in 1990. The other condition for the Kyoto Protocol to enter into force (ratification by 55 signatories) had already been satisfied in 2002. Australia ultimately ratified the Kyoto Protocol in 2007 during the COP-13 after the election of Prime Minister Kevin Rudd.

Commitments under the Kyoto Protocol and Kyoto Mechanisms
During the first commitment period (2008–12), countries with commitments under the Kyoto Protocol (38 industrialized countries and the European Community) committed to reduce their aggregate CO_2e emissions[3] on average by 5 per cent against 1990 levels. These countries are also referred to as Annex B Parties (Annex I Parties not including Turkey and Belarus). The target was not the same for all countries, but was set relative to each Party's GHG emissions in a specific reference year, also referred to as the base year. For most countries, the base year was defined as 1990 (UNFCCC, 2007b). The allowable level of GHG emissions is referred to as the 'Party's assigned amount' over the 2008–12 commitment period, denominated in tons of CO_2 equivalent emissions (CO_2e) which are informally referred to as 'Kyoto units'. Annex B Parties were thus required to ensure that they did not exceed their assigned amount over the first commitment period from 2008–12.

Parties were asked to implement domestic measures to reduce their GHG emissions. These measures included offsetting emissions by increasing so-called 'carbon sinks' (that is, projects or activities that absorb more carbon than they release) in land-use, land-use change and forestry (LULUCF). In order to verify emissions, Annex B Parties needed

---

[2]   The US signed the Kyoto Protocol under the Clinton Administration, but the Kyoto Protocol was never presented the Protocol to the Senate for ratification. This was attributed by many to the Byrd-Hagel Resolution, but the reason has been subject to debate.

[3] CO_2e is a quantity that describes, for a given GHG, the amount of CO_2 that would have the same global warming potential when measured over a specified timescale (generally 100 years).

to have in place a national system for estimating the anthropogenic GHG emissions by sources and removals by sinks at least one year prior to the start of the first commitment period (2008–12) (UNFCCC, 2007b). In addition, the Kyoto Protocol introduced three mechanisms to Parties for the purposes of cutting their emissions or enhancing their carbon sinks more cheaply abroad than within their own territory. These mechanisms, referred to as Joint Implementation (JI), the Clean Development Mechanism (CDM), and Emissions Trading (ET), take place in addition to domestic action. To be eligible to participate in the mechanisms, Parties under Kyoto must meet certain eligibility criteria that are based on the methodological and reporting requirements related to GHG inventories (UNFCCC, 2007b).

International Emissions Trading Emissions trading (Kyoto Protocol, Article 17) allows countries with an emission-reduction or emission-limitation commitment under the Kyoto Protocol (Annex B Parties) to trade their emissions. In other words, countries that have emission units to spare (namely, countries whose assigned amount is greater than their emissions 'used') are allowed to sell this excess capacity to countries that are over their national target. Emissions trading thus created a new commodity in the form of emission reductions or removals. Since carbon dioxide is the principal GHG, the trading of emissions is often referred to as trading in carbon. The market associated with the trading is known as the 'carbon market' (UNFCCC, 2013i).

Clean Development Mechanism The CDM (Kyoto Protocol, Article 12) allows any Annex B Party to implement emission-reduction projects in developing countries.[4] These CDM projects can generate certified emission reduction (CER) credits (equivalent to one ton of CO_2) that can then be counted towards meeting Kyoto targets. A project activity might involve, for example, a rural electrification project using solar panels (UNFCCC, 2013i). The CDM is meant to lead to emission reductions while stimulating sustainable development, and giving Annex B Parties some flexibility in meeting their emission targets (UNFCCC, 2013i).

Joint Implementation The JI (Kyoto Protocol, Article 6) allows any Annex B Party to invest in approved emission reduction or emission removal programmes or projects in another Annex B country with the

---

[4] Project must meet certain requirements. CDM projects must provide emission reductions that are real, measurable, and verifiable, and fulfill the criterion of 'additionality' (that is, emission reductions need to be additional, to reductions that would have occurred without the project).

objective of obtaining emission reduction units (ERUs) from a programme or project that can be counted towards meeting the Kyoto target.
JI is meant to provide a flexible and cost-efficient mechanism for
countries to achieve part of their Kyoto commitments, while the host
country benefits from technology transfer and foreign investment
(UNFCCC, 2013i).

Towards a Future of Kyoto

In accordance with Kyoto Protocol requirements, Parties to the Kyoto
Protocol launched negotiations on the next phase of the Kyoto Protocol at
the first Meeting of the Parties to the Kyoto Protocol in 2005 in Montreal
(CMP-1, held in conjunction with COP-11). The negotiations took place
within the so-called 'Ad Hoc Working Group on Further Commitments
for Annex I Parties under the Kyoto Protocol', also referred to as the
AWG-KP or Kyoto Protocol Track (UNFCCC, 2013a). However, it took
several years of intense international negotiations until progress was
made towards achieving a post-2012 (post-Kyoto) agreement, and a
long-term climate agreement is still not implemented. Important docu-
ments along the way include the Bali Roadmap, the Copenhagen Accord,
the Cancun Agreements, and the Durban Platform for Enhanced Action.

The Bali Roadmap (COP-13, 2007) The main focus of COP-13 in Bali
was to find a successor agreement to the Kyoto Protocol, and to make
progress on long-term action on climate change. Parties agreed on the
so-called Bali Roadmap, which laid out the way towards a post-Kyoto
(post-2012) outcome. The Bali Roadmap consisted of the outcomes of
the negotiations of two 'tracks', the so-called Bali Action Plan Track, and
the AWG-KP (or Kyoto Protocol) Track. The central task of the
AWG-KP Track was to negotiate the emission reduction commitments of
industrialized countries for after the end of the Kyoto Protocol's first
commitment period (UNFCCC, 2013b). It was anticipated that negotiations of a post-2012 agreement would conclude at COP-15 in Copenhagen at the end of 2009, thus giving Parties time to ratify the treaty for
it to take effect at the end of 2012 and to follow on from commitments
under Kyoto.

One of the key outcomes under the Bali Roadmap was the Bali Action
Plan, which considered how to significantly upscale policy responses to
climate change, including adaptation. Key outcomes of the Bali Action
Plan included:

- To 'launch a comprehensive process to enable the full, effective and
  sustained implementation of the Convention through long-term

cooperative action, now, up to and beyond 2012, in order to reach an agreed outcome and adopt a decision at its fifteenth session' (United Nations Conference of Parties, 2007). This process encompassed the following pillars (Barrett, 2008): (1) a shared vision for long-term cooperative action on climate change, including a long-term global goal for emission reductions, (2) enhanced national and international action on climate change mitigation, (3) enhanced action on adaptation, (4) enhanced action on technology development and transfer to support action on mitigation and adaptation, and (5) enhanced action on the provision of financial resources and investment to support action on mitigation, adaptation and technology cooperation.

- To establish a subsidiary body under the Convention, the 'Ad Hoc Working Group on Long-term Cooperative Action under the Convention' (also referred to as the AWG-LCA or the Convention Track). This negotiating track was set up as a track running parallel to the Kyoto Track, and was meant to be a 'broader' track that included the United States (as a non-Kyoto party) as well as developing countries. The role of the AWG-LCA Track was to address issues relating to new mechanisms for technology, finance, mitigation and adaptation, and to also present the outcome of its work to the Conference of the Parties for adoption at COP-15 in 2009. Together, the Kyoto Track and the Convention Track were thus working under the same deadline and meant to be leading to a comprehensive global climate regime.

In line with the IPCC Fourth Assessment Report (AR4), the Bali Action Plan emphasized that global warming is 'unequivocal' and that a 'delay in reducing emissions significantly constrains opportunities to achieve lower stabilization levels and increases the risk of more severe climate change impacts' (United Nations Conference of Parties, 2007). While a footnote in the Bali Action Plan text referred to the urgency of addressing climate change based on the AR4 (which includes a goal of halving global emissions by 2050 compared to 2000 levels), neither the Bali Action Plan nor the Bali Roadmap specified or introduced any new emission targets. Developed countries were urged to consider 'measurable, reportable and verifiable nationally appropriate mitigation commitments or actions' for reducing their GHG emissions, while developing countries were urged to undertake 'nationally appropriate mitigation actions' (NAMAs) supported by technology, financing and capacity-building. This left key uncertainties around issues such as the level of emission reductions required from both developed and developing nations in the longer term and the timeframes and mechanisms by which these emission reductions can/should be achieved, as well as the implications for specific countries and economic sectors.

The Bali Roadmap also included discussions of new policy areas, such as the reduction of emissions from deforestation in developing countries (REDD) through policy approaches and possible financial support, leading to the launch of the UN-REDD Programme in 2008. In addition, the Bali Roadmap suggested the urgent implementation of adaptation actions. The Bali Roadmap also initiated steps for setting up mechanisms to encourage the development and transfer of technology, especially from developed to developing countries to help them to reduce or avoid carbon emissions and to adapt to the impacts of climate change. Other key outcomes of the Bali Conference included the launch of the Adaptation Fund. The Adaptation Fund was set up under the Kyoto Protocol in 2007 to help developing country Parties to deal with climate change. Funding is sourced by a levy of 2 per cent on certified emission reductions (CERs) on projects under the Kyoto Protocol's CDM. The first projects were approved in 2012.[5]

The Copenhagen Accord (COP-15, 2009) At COP-15 and among stalling negotiations, the so-called Copenhagen Accord was drafted by the United States, and the BASIC countries Brazil, South Africa, India and China. The COP 'took note' of the Accord, but it did not become the binding successor for Kyoto that many had hoped for. It was thus more a political than a legal agreement (PEW Center on Global Climate Change, 2009). The Copenhagen Accord outlined several key points on which the views of the drafting governments converged. One key point was the goal to limit the global average temperature increase to a maximum of 2°C above pre-industrial levels, alongside the recognition that ambitious cuts in global GHG emissions are essential to maintain the chances to achieve this goal. The Copenhagen Accord, however, did not offer details on how to achieve such cuts, and there was no agreement among countries on how to implement stringent emission reductions in practical terms. The Accord also included a reference to consider limiting the temperature increase to a maximum of 1.5°C. The inclusion of this temperature target was a key demand made by developing countries (UNFCCC, 2013e).

Other key points of the Copenhagen Accord included pledges by developed countries to fund actions on GHG emission reductions (which

---

[5] The progress of funded initiatives has been made available by The Germanwatch Adaptation Fund Project Tracker, accessed 24 September 2014, available at: http://af-network.org/4889.

had to be entered by 31 January 2010), but also actions on adaptation to climate change in developing countries. The Copenhagen Accord proposed mechanisms for developed countries to 'provide adequate, predictable and sustainable financial resources, technology and capacity-building to support the implementation of adaptation action in developing countries'. Developing countries were promised short-term funds of US$30 billion for the period 2010–12 to finance adaptation and mitigation measures, especially in the least developed countries and those countries most affected by climate change. Developed countries promised to mobilize long-term finance of a further US$100 billion per year by 2020 from 'a wide variety of sources' to support developing countries' needs.

The Copenhagen Accord called for the establishment of a High-Level Panel under the COP to examine how to meet the 2020 finance goal, a Copenhagen Green Climate Fund (as an operating entity under the High-Level Panel), and a Technology Mechanism to accelerate technology development and transfer for both adaptation and mitigation (UNFCCC, 2013e). In addition, the Copenhagen Accord recognized the importance of REDD and the 'need to enhance removals or greenhouse gas emission by forests', and called for 'the immediate establishment of a mechanism including REDD-plus to enable the mobilization of financial resources from developed countries'.[6] Despite its weaknesses, the Copenhagen Accord received support from 141 countries.[7] More than 90 Annex I and non-Annex I Parties, including all major emitters, entered pledges to reduce their emissions, or constrain their growth, mainly up to 2020.[8] However, as the Copenhagen Accord was a non-binding agreement, these pledges did not become binding emissions targets under Kyoto.

The Cancun Agreements (COP-16, 2010) Delegates faced high pressure in Cancun, Mexico, to make progress towards negotiating an international approach to addressing climate change in order to avoid a sidelining and failure of the UNFCCC process (World Resources Institute, 2010). This time, however, and in contrast to Copenhagen, the

---

[6]   REDD-plus refers to co-benefits of REDD, such as conservation, sustainable management of forests and the enhancement of forest carbon stocks in developing countries. The concept was first introduced by the Convention's Subsidiary Body for Scientific and Technological Advice (SBSTA) during its meeting in December 2008.

[7]   As of September 2013.

[8]   Accessed 21 September 2014, available at http://unfccc.int/home/items/5264.php and http://unfccc.int/home/items/5264.php.

negotiation process was facilitated as mitigation targets and pledges submitted by individual countries as part of the Copenhagen Accord were not under negotiation. Rather, the focus was largely on negotiating how the non-binding pledges submitted under the Copenhagen Accord, including pledges from the US and China, would be reflected in the UN decisions and in official UN documentation (PEW Center on Global Climate Change, 2010). A coalition, the so-called Cartagena Dialogue, had formed since Copenhagen among a number of developing and developed countries to foster South-North cooperation, and had already made progress towards negotiating compromises prior to COP-16 (World Resources Institute, 2010).[9]

One of the important outcomes of COP-16 was that central elements of the Copenhagen Accord became incorporated into the formal UNFCCC process (PEW Center on Global Climate Change, 2010). Formal recognition was given to the goal to limit the global average temperature increase to a maximum of 2°C above pre-industrial levels, with the agreement to consider a limit to 1.5°C. Negotiations under the AWG-LCA Track resulted in decisions which were based on, and incorporated the pillars of the Bali Action Plan from 2007: a shared vision for long-term cooperative action; enhanced action on adaptation (giving rise to the Cancun Adaptation Framework and an Adaptation Committee, discussed below); enhanced action on mitigation (through quantified economy-wide emission reduction targets by developed countries, and NAMAs by developing countries); as well as finance, technology, and capacity-building (including technology development and transfer).

Progress was also made on policy approaches and incentives for reducing emissions from deforestation and forest degradation in developing countries (the REDD/REDD-plus policy mechanism) discussed under the Bali Action Plan and the Copenhagen Accord. The Cancun Agreements outlined a phased approach to strengthen efforts by developing countries, asking them to develop a national strategy or action plan, a national forest reference emission level and/or forest reference level, as well as transparent national forest monitoring systems. However, financing considerations were not fully resolved (PEW Center on Global Climate Change, 2010).

---

[9]   There is no formal membership to the Dialogue, and thus no membership list with countries that actively participate in the Dialogue. For an informal list: see Pineda, C. (2012) *The Cartagena Dialogue: A Bridge to the North-South Divide?* Brown University, Center for Environmental Studies.

The mechanisms and actions envisioned under the Copenhagen Accord were incorporated as well. Developed countries were asked to formalize their commitments to raise funds 'approaching USD 30 billion for the period 2010–2012' (the 'fast start' period) to support developing countries' climate efforts. Parties agreed to establish the Green Climate Fund which was initially proposed in Copenhagen and intended to be the centerpiece of efforts to raise climate finance to support developing countries in adapting to climate change. The World Bank was designated as the interim trustee of the fund (PEW Center on Global Climate Change, 2010). The proposed Technology Mechanism also received further backing.

Decisions on the future of the Kyoto Protocol and a second commitment period, however, were effectively deferred until the COP-17 in Durban (see below). Many countries wanted a successor to Kyoto negotiated during Cancun, as the first commitment period was going to end in 2012. While Cancun did not reach agreement on a second commitment period, a number of decisions demonstrated progress under the AWG-KP (Kyoto Protocol) Track. The Cancun Agreements under the Kyoto Protocol Track recognized:

> that the contribution of Working Group III to the Fourth Assessment Report of the Intergovernmental Panel on Climate Change, *Climate Change 2007: Mitigation of Climate Change*, indicates that achieving the lowest levels assessed by the Intergovernmental Panel on Climate Change to date and its corresponding potential damage limitation would require Annex I Parties as a group to reduce emissions in a range of 25–40 per cent below 1990 levels by 2020, through means that may be available to these Parties to reach their emission reduction targets.

The Agreements under the Kyoto Protocol Track took note of the targets of Annex I Parties, thus helping to formalize the targets in the lead-up to a second commitment period. Countries agreed that further work was needed for converting these targets into binding commitments under the Kyoto Protocol. Despite the goal to limit global temperature increase to a maximum of 2°C above pre-industrial level, all pledges put forward by governments only amounted to about 60 per cent of the emission reductions needed for a 50 per cent chance of keeping temperatures below the 2°C target (UNFCCC, 2013d). The Cancun Agreements also included agreements regarding the continuation of the Kyoto Mechanisms (ET and JI) for meeting Annex I targets. Carbon capture and storage (CCS) was approved as an eligible project type under the Kyoto Protocol's CDM, subject to safety and technical considerations. There was also agreement that the negotiations in regards to a successor of the

Kyoto Protocol to leave no gap between the first and second commitment period of the Kyoto Protocol (UNFCCC, 2013d; World Resources Institute, 2010).

The Durban Platform for Enhanced Action (COP-17, 2011) At COP-17, and with the expiration of Kyoto's initial targets being imminent, countries needed to agree on a successor agreement to Kyoto. The negotiations resulted in agreement on the following broad areas:

1. **A second commitment period of the Kyoto Protocol** Under the AWG-KP (Kyoto Protocol) Track, a second commitment period of the Kyoto Protocol was agreed from 1 January 2013 onwards. The European Union and a few developed countries agreed to take a second set of binding emission targets, but only if other Parties agreed to launch a new round of negotiations toward a comprehensive successor agreement starting in 2020 (see below). Parties declared their intention to convert their emission reduction pledges into quantified emission limitation and reduction objectives (QELROs) in an amendment to the Kyoto Protocol, to be adopted at the next Meeting of the Parties to the Kyoto Protocol (namely, CMP-8, held in conjunction with COP-18), with provisions for immediate implementation (PEW Center on Global Climate Change, 2011).

2. **The launch of a new platform of negotiations** It was decided that the post-2020 talks about a successor agreement would be conducted by a new working group, the Ad Hoc Working Group on the Durban Platform for Enhanced Action, with a deadline of 2015. The objective for this working group is to deliver a new and universal GHG reduction protocol, legal instrument, or other outcome with legal force by 2015 for the period beyond 2020. Whether or not the new agreement would be legally binding and apply to both developed and developing countries were major points for discussion during the negotiations. The final compromise was that the Parties agreed to 'develop a protocol, another legal instrument or an agreed outcome with legal force under the Convention applicable to all parties' – which implies (but does not explicitly mandate) that the post-2020 agreement will be legally binding (PEW Center on Global Climate Change, 2011). Some commentators regarded this formulation as a major turning point in the UNFCCC negotiations, as it avoided a strict differentiation between developed and developing countries which was reflected in the Kyoto Protocol (with legally binding emission targets for the developed countries and no new commitments for developing

countries) (Center for Climate and Energy Solutions, 2013). COP-17 increased the warning around the threats caused by climate change and concluded that: 'climate change represents an urgent and potentially irreversible threat to human societies and the planet and thus requires to be urgently addressed … with a view to accelerating the reduction of global greenhouse gas emissions'.

3.   **Conclusion in 2012 of existing broad-based stream of negotiations** In addition to the agreement under the AWG-KP Track, a decision was also reached to conclude the work of the existing AWG-LCA Track within 2012. A major outcome in Durban was the formal launch of the Green Climate Fund to support adaptation and mitigation outcomes in developing countries; however, there was still no indication as to when developed countries would make their promised contributions (PEW Center on Global Climate Change, 2011). It was also agreed at Durban to make the Technology Mechanism fully operational by 2012 'to promote and enhance the research, development, deployment and diffusion of environmentally sound technologies in support of action on mitigation and adaptation in developing countries' (UNFCCC, 2011c). Additionally, procedures were adopted to make existing national emission reduction (or limitation) plans more transparent. A number of other issues remained unresolved, and were deferred until the following year. These included a global goal for emission reductions by 2050, the scope and means of a review of the 2°C target, as well as the overall progress towards achieving it (Center for Climate and Energy Solutions, 2011).

The Doha Amendment to the Kyoto Protocol (COP-18, 2012)   The UN Climate Change Conference in Doha, Qatar, resulted in a package of decisions referred to as the Doha Climate Gateway. The most significant achievement of the COP-18 Conference was the adoption of an amendment to the Kyoto Protocol, which – temporarily – extended its life and established a second round of binding GHG emission targets for Europe, Australia and some other developed countries[10] (Center for Climate and Energy Solutions, 2012). As the amendment had to be ratified by three-quarters of the Parties to legally enter into force, those countries with targets agreed to either provisionally apply the amendment, or to implement their new commitments from 1 January 2013 onwards. The

---

[10]   Belarus, Iceland, Kazakhstan, Liechtenstein, Luxembourg, Monaco, Norway, Switzerland and Ukraine.

amendment also concluded the negotiations that began in 2005 in Montreal under the AWG-KP Track (Center for Climate and Energy Solutions, 2012). A work programme was agreed upon to further investigate new market-based mechanisms – such as nationally-administered or bilateral offset programmes – and their role in supporting countries to meet their mitigation targets (UNFCCC, 2013g).

Parties also took final decisions under the parallel AWG-LCA Track, that was launched in 2007 in Bali. Parties had overall very different views of what still needed to be done to fulfil the AWG-LCA's mandate under the 2007 Bali Action Plan. Developed countries emphasized the accomplishment in Cancún and Durban, including new mechanisms and bodies on technology, finance, mitigation and adaptation, as well as improved mechanisms for transparency. Developing countries, on the other hand, argued that a range of issues remained unresolved and that further decisions would be required, especially in regards to a long-term global emission goal (Center for Climate and Energy Solutions, 2012). Even though issues surrounding some long-term outcomes remained unresolved, the UNFCCC brought the two long-standing negotiating tracks to a close. Instead, it advanced a single new track, called the Durban Platform for Enhanced Action, for negotiating a comprehensive legal agreement by 2015. The Durban Platform calls for a universal climate agreement with legal force 'applicable to all' to be established by 2015, starting in 2020. However, the Parties only managed to agree on procedural steps and an overall loose process for the next years of negotiations until COP-21 in 2015, when the Durban Platform talks are to conclude (Center for Climate and Energy Solutions, 2012).

The outcomes in Doha did not come easy. As several tracks were being negotiated simultaneously, the talks were procedurally quite complex. Small island countries and the least developed countries (those most vulnerable to climate change) pushed for a strengthening of both existing emission pledges and financial commitments by developed countries. In particular, the countries hoped for stronger commitments following President Obama's re-election and the devastation caused by Hurricane Sandy. The major concession to these countries was the launch of a process to consider establishment of a new mechanism for addressing 'loss and damage' from climate impacts, including weather extremes and slow-onset impacts such as sea-level rise. The United States was success-ful in avoiding the inclusion of any language that could suggest that such a mechanism would provide countries with some form of direct compen-sation for losses (Center for Climate and Energy Solutions, 2012).

COP-19 in Warsaw, 2013 COP-19 marked the midpoint in the post-2020 talks launched at Durban. A central focus of COP-19 was to define a clearer agenda for the final two years of the negotiations of the Ad Hoc Working Group on the Durban Platform for Enhanced Action. Parties set a loose timeline for proposing their 'intended nationally determined contributions' by the 2015 deadline for those 'ready to do so'. At COP-19, there were major demands from developing countries for increased climate finance, and for establishing the new mechanism to address 'loss and damage' from climate impacts that was initially brought forward at COP-18 at Doha. Addressing 'loss and damage' took on new prominence for many developing countries as Typhoon Haiyan struck the Philippines just days before the conference. Overall, however, the Warsaw COP-19, Conference had the least impact compared to past conferences that resulted in more substantial outcomes, such as decisions to negotiate a successor agreement to the Kyoto Protocol (Center for Climate and Energy Solutions, 2013). The negotiations resulted in the following key outcomes (UNFCCC, 2013g):

1. **Ad Hoc Working Group on the Durban Platform for Enhanced Action** At COP-19, there was not a great deal of progress on a pathway towards a climate agreement in 2015. The adopted decision invited 'all Parties to initiate or intensify domestic preparations for their intended nationally determined contributions ... by the first quarter of 2015 by those Parties ready to do so', and to communicate them 'well in advance' of COP-21 in Paris. However, this decision called for a further decision at COP-20 to define the information that needs to be provided when putting these contributions forward, so that Parties can better understand and assess each other's proposals. The decision also urged developed countries to provide support to developing countries in order to enable them to develop their intended contributions.

2. **Loss and damage** Some members of the Alliance of Small Island States (AOSIS) came to Warsaw to push for a 'compensation' mechanism for 'loss and damage' resulting from extreme events and slow-onset impacts such as sea-level rise. The proposal received strong resistance from the United States and other developed countries, and was thus not further pursued at COP-19. However, Parties agreed to establish the 'Warsaw international mechanism for loss and damage associated with climate change impacts' to share data, information and best practices among relevant stakeholders, and to enhance action and support, including finance, technology and capacity building to address 'loss and

damage'. The United States successfully insisted that the new Warsaw international mechanism became a workstream as part of the existing Cancún Adaptation Framework (and not a new 'pillar' of the Convention, as suggested by AOSIS) with an executive committee reporting directly to the COP. However, Parties agreed to revisit the Warsaw international mechanism and its structure at COP-22 in 2016.

3.   **Finance** As part of the Copenhagen and Cancún agreements, developed countries promised short-term funds of US$30 billion for the period 2010–12 to finance adaptation and mitigation measures especially for the least developed countries, and those countries most affected by climate change. In addition, developed countries promised to mobilize long-term finance of a further US$100 billion per year by 2020 from 'a wide variety of sources' to support developing countries' needs (see relevant sections above). Developing countries voiced concerns over the lack of progress in securing the funding in Warsaw, and argued that an interim goal of $70 billion by 2016 should be set. This suggestion did not receive any backing from developed countries, which only agreed to submit new biennial reports outlining their strategies for scaling-up climate finance. However, COP-19 introduced a biennial ministerial dialogue on climate finance running from 2014 to 2020. The COP made progress towards completing arrangements with the Green Climate Fund to enable it to mobilize funding and commence its operations.

4.   **Other areas** Parties made some progress on REDD-plus and adopted guidelines for the development of 'reference levels' against which the efforts of individual countries to reduce deforestation will be measured. Norway, the UK and the United States pledged a total of $280 million for REDD-plus efforts. Little progress was made toward discussing new market-based mechanisms under the UNFCCC, in part because some Parties were strongly opposed to market-based approaches. The discussions remained bogged down in the Subsidiary Body on Scientific and Technological Advice (SBSTA) and never reached the COP, but will be taken up again at COP-20 (Center for Climate and Energy Solutions, 2013).

The Mandate for Adaptation

As policy approaches have largely focused on climate change mitigation, a dichotomy started to emerge between the concepts of adaptation and mitigation. This is reflected not only in relation to law-making, but also in the treatment of the two issues in policy and academic debates on the merits of adaptation and mitigation (Parry, 2009; Tol, 2005). Since the early 2000s, however, political perceptions on adaptation have slowly begun to change. Since COP-7, held in Marrakech in 2001, greater emphasis has been placed on recognizing and responding to the vulnerability of countries to the adverse impacts of climate change, with a particular focus on developing countries. The subsequent releases of the IPCC's *Fourth* and *Fifth Assessment Reports* (IPCC, 2007a; 2013) strongly re-emphasized the need for adaptation, and adaptation started to emerge as a policy issue alongside mitigation.

In examining the UNFCCC process, Schipper (2006) concluded that a number of specific reasons can be identified for why adaptation has not received as much attention as mitigation early on in international negotiations. A first reason is that adaptation was regarded as a 'defeatist' approach – an early focus on adaptation would have meant acknowledging that climate change is impactful, that mitigation alone is not sufficient, and that adjustments are required that go beyond 'business-as-usual'. Therefore, a focus on adaptation was seen as not constructive. A second reason is that planning for adaptation would have been seen as an admission that climate change was definitely occurring, at a time when there was great skepticism about the scientific certainty around global warming. The Global Climate Coalition, for instance, a lobbying group that was active from 1997 to 2002 and represented large (mostly US-based) organizations such as Exxon and Shell, questioned the science behind the UNFCCC.[11]

Furthermore, while the need to adapt to the adverse effects of climate change is mentioned in the text of the UNFCCC treaty (United Nations, 1992), the UNFCCC has no article solely dedicated to the topic of adaptation, and there are only six references to adaptation in the Convention text. The Convention and subsequent agreements do not contain a clear definition of the concept or any additional guidance on how adaptation should be facilitated. Negotiators have argued that this has left insufficient opportunities to address adaptation under the UNFCCC and that – as a result – the main focus of the UNFCCC negotiations rested on efforts towards the mitigation of GHG emissions (Schipper, 2006).

---

[11]   The group eventually fell apart when several members left, recognizing that firms need to do more than ignore climate change and be active to avoid risks, and embrace opportunities.

There is also evidence that, within the highly politicized early debate on climate change and accountability, developed countries were concerned that any action on adaptation would have been seen as 'an implicit assumption of responsibility for causing climate change' (Sands, 1992: 275). Any admission by developed countries that their economic development was a key driver behind climate change could have possibly led to subsequent questions around financial responsibilities, liability and compensation. In addition, developing countries did not want to take the debate away from commitments to emission reduction targets by developed countries, thus avoiding a further discussion of adaptation targets. The science around climate change was also still in its infancy and had not established a clear connection between gradual changes and changes in the frequency and/or intensity of weather extremes, thus making adaptation seem a less urgent, imminently-required action. Overall, mitigation was considered as more beneficial and cost-effective to avoid negative impacts in the first place (Schipper, 2006).

Adaptation as an emerging topic within the UNFCCC

Within policy negotiations, adaptation has been a primary concern to small island developing states which are concerned about adverse effects such as sea level rise. These states organized themselves within the Alliance of Small Island States (AOSIS) and proposed a set of tasks on adaptation for the UNFCCC, highlighting in particular the need for adaptation funding. However, after the UNFCCC entered into force in 1994, the main focus shifted towards mitigation, and towards negotiating the Kyoto Protocol as an instrument for reducing GHG emissions. Following its adoption in 1997, Parties then agreed that a number of issues needed to be addressed before the Kyoto Protocol could be implemented as set out in the 1998 Buenos Aires Plan of Action. The negotiations were meant to culminate at COP-6 in 2000; however, due to failure in negotiations, it was not until COP-11 in 2005 that the Marrakech Accords were adopted as a set of rules for the implementation of Kyoto. This period in between the adoption of the Kyoto Protocol (1997) and it entering into force (2005) then finally opened up some space for adaptation to emerge in policy discussions (Schipper, 2006).

It soon became clear that the inclusion and implementation of adaptation objectives was not as easy as anticipated (Schipper, 2006). The UNFCCC and a subsequent decision adopted at COP-1 addressed the need to adapt and fund adaptation activities, especially in highly vulnerable developing countries. This eventually led to the agreement to establish a Least Developed Countries Expert Group and Least Developed Countries Fund in 2001 under the Global Environment Facility

(GEF) as part of the Marrakech Accords to support the least developed countries in preparing and implementing National Adaptation Pro-grammes of Actions (NAPAs) – with a focus on urgent and immediate needs. The Least Developed Countries Expert Group became the first body supporting adaptation, and the NAPAs became the first workstream on adaptation. The NAPAs covered activities such as the prevention of loss of life and livelihood, health, food security, water availability and essential infrastructure, among others. A Special Climate Change Fund (SCCF) was established in 2001 (under the Global Environment Facility) to finance projects, including those related to adaptation, technology transfer, energy, transport, industry, agriculture, forestry and waste man-agement, as well as economic diversification in developing countries that are highly dependent on income generated from energy-intensive fossil fuel production, processing, export and/or consumption.

At COP-9, Parties requested the initiation of a more comprehensive work programme on adaptation. At COP-10 in 2004, the SBSTA received a mandate to develop a five-year work programme on the scientific, technical and socio-economic aspects of impacts, vulnerability and adaption to climate change (2005–10) with a focus on the dissemination of knowledge and information on adaptation. This program was renamed at COP-12 in 2006 to the 'Nairobi work programme on impacts, vulnerability and adaptation to climate change' (or the Nairobi Work Programme on Adaptation), and became the second workstream on adaptation. The goal of establishing the work programme was to enhance the Parties' understanding and assessment of climate change impacts, vulnerabilities, and adaptation options, and to enable them to make informed decisions on adaptation measures. Specifically, the programme was set up with the following objectives:

1. Improve capacity at international, regional, national, sectoral and local levels to further identify and understand impacts, vulnerability and adaptation responses, in order to effectively select and imple-ment practical, efficient and high priority adaptation actions.
2. Enhance and improve the level and amount of information and advice on the scientific, technical and socio-economic aspects of impacts, vulnerability and adaptation.
3. Enhance the degree of dissemination and utilization of knowledge from practical adaptation activities.
4. Enhance cooperation among Parties, relevant organizations, busi-ness, civil society and decision-makers to advance their ability to manage climate change risks.

5.   Enhance the integration of adaptation into sustainable development plans (UNFCCC Secretariat, n.d.). This includes work on impact and vulnerability assessments, as well as adaptation assessment, planning and actions (UNFCCC, 2007b, n.d.).

At COP-17, the SBSTA was asked to reconsider the work areas of the Nairobi Work Programme with a view to making recommendations to the COP at its 19th Session on future aims and objectives of the programme. Under the Nairobi Work Programme, the Private Sector Initiative (PSI) was set up to catalyze the engagement of the private sector in adaptation and to offer a platform for businesses to acquire and share knowledge about adaptation. Private sector organizations can join the PSI to demonstrate their progress with assessing vulnerabilities and opportunities, and with developing an adaptation strategy. An online database was created to feature case studies of adaptation practices and profitable climate change adaptation activities undertaken by private companies (sometimes in partnership with NGOs or the public sector) from a wide range of regions and sectors.[12] The case study database is intended as a tool to showcase and exchange best practices.

Parties to the UNFCCC made further commitments to adaptation under the Cancun Agreements and adopted the Cancun Adaptation Framework at COP-16 in Cancun in 2010. The Cancun Adaptation Framework was the result of three years of negotiations on adaptation under the AWG-LCA Track, following the adoption of the Bali Action Plan at COP-13 in 2007. The Bali Action Plan sought to enable 'the full, effective and sustained implementation of the Convention through long-term co-operative action, now, up to and beyond 2012' (United Nations Conference of Parties, 2007). Parties affirmed in the Cancun Agreements that adaptation must be addressed with the same level of priority as mitigation. Furthermore, Parties established that international cooperation and enhanced action on adaptation are urgently required to enable and support the implementation of adaptation action with the aim to reduce vulnerability and build resilience in developing country Parties (UNFCCC, 2013c). Specifically, Parties under the Cancun Adaptation Framework agreed to:

---

[12]   Accessed 25 September 2014, available at: http://unfccc.int/adaptation/workstreams/nairobi_work_programme/items/6547.php

1.   plan, prioritize and implement adaptation actions;
2.   conduct impact, vulnerability and adaptation assessments, including financial needs assessments and an evaluation of economic, social and environmental adaptation options;
3.   strengthen institutional capacities for adaptation;
4.   build resilience of socio-economic and ecological systems;
5.   enhance climate change-related disaster risk reduction strategies, including early-warning systems, risk management and insurance;
6.   enhance understanding, cooperation and coordination with regards to climate change-induced displacement, migration and planned relocation;
7.   research, develop, deploy and transfer technologies and processes for adaptation;
8.   strengthen information systems and public awareness; and
9.   improve climate research, observation and modelling (UNFCCC, 2011b).

Under the Cancun Adaptation Framework, a process was established for least-developed countries to formulate and implement National Adaptation Plans (NAPs). These NAPs became a third workstream, and built on existing NAPAs which were already established to report on and plan for adaptation needs, but with a focus on identifying medium- and long-term adaptation needs. The COP-18, in decision 12/CP.18, provided direction to the Global Environment Facility to enable activities for the preparation of National Adaptation Plans through the Least Developed Countries Fund while maintaining support for the first workstream on NAPAs. Parties other than least developed countries were invited to support the national adaptation plans with funding through the Special Climate Change Fund (SCCF). Under the Cancun Adaptation Framework, a work programme was established to consider approaches for addressing 'loss and damage' associated with climate change impacts in developing countries. COP-19 than established the Warsaw international mechanism for loss and damage associated with climate change impacts (see decision 2/CP.19). The initial meeting of the Executive Committee of the Warsaw international mechanism for loss and damage was scheduled for March 2014. This program became a fourth (and contentious) workstream. An Adaptation Committee was established to support the implementation of enhanced actions on adaptation. The Adaptation Committee became the second body supporting adaptation.

Non-UNFCCC adaptation programmes

In addition to the UNFCCC actions on adaptation to climate change, there are a significant number of initiatives occurring outside UNFCCC-driven adaptation programs. Many international non-governmental organizations (NGOs), education institutions, public-private partnerships, consortia, and even for-profit organizations have programs and/or funding resources in place, which are often directly or indirectly addressing development and adaptation needs. Examples include the 2030 Water Resources Group, a public-private platform for collaboration on water issues which seeks to mobilize stakeholders from the public and private sector (including SABMiller, Nestlé, Coca-Cola), civil society, academia and financing institutions to build initiatives in local markets that help with the adaptation to climate change and water scarcity.

CASE STUDY: CORPORATE ENGAGEMENT IN INTERNAL AND EXTERNAL POLICYMAKING FOR ADAPTATION AND RESILIENCE

Eskom, a South African electricity supply company, has recently undertaken adaptation actions to deal with the threat of water scarcity. For Eskom, this threat has become one of the most important risks of climate change as the company is a major user of South Africa's fresh water resources as it uses water as a coolant in its coal-fired power plants (United Nations Global Compact and United Nations Environment Programme, 2012).

Eskom has been developing ways to deal with the risk of water scarcity at both a plant and executive level, including research and analysis, and the development of a specific adaptation strategy to manage risk and build resilience. In 2010, Eskom engaged in a climate change vulnerability assessment process, the purpose of which was to analyze historical and current weather conditions, climate variability and extreme weather events, and to ascertain the impacts of these variables on Eskom's business (United Nations Global Compact and United Nations Environment Programme, 2012).

This process led to a comprehensive adaptation strategy in 2011–12. The objectives of this strategy are to:

- Ensure delivery of coordinated and timely climate information within the company.
- Use the company's existing tools to integrate needs assessment, implementation, evaluation, and reporting on climate change adaptation within the company.
- Define and quantify weather, climate variability and other long-term impacts of climate change on Eskom's business.
- Invest in climate change research.

- Integrate climate change adaptation into the business' stakeholder communications and reputation-building initiatives.

Eskom started to establish a consolidated business position on the potential impacts of climate change on all aspects of business activities (United Nations Global Compact and United Nations Environment Programme, 2012). The company became actively involved in adaptation policy-making, providing written inputs to the South African government's National Climate Change Response White paper and engaging in international policy dialogues on climate change adaptation (United Nations Global Compact and United Nations Environment Programme, 2012).

Financing adaptation

Within the UNFCCC, the main sources for financing adaptation are those outlined above: The Least Developed Countries Fund (LDCF) to provide financial support for the preparation and implementation of NAPAs, and the Special Climate Change Fund (SCCF) to finance projects related to, inter alia, adaptation. Both these funds were established under the Global Environment Facility. The Adaptation Fund was set up under the Kyoto Protocol to help developing country Parties to deal with climate change and to provide financial assistance to adaptation projects in developing country Parties (funding is raised through a levy on 'carbon offset' revenues through the CDM). At COP-17, Parties agreed to establish the Green Climate Fund to provide support for adaptation and mitigation outcomes in developing countries; however, there was still no indication as to when developed countries would make their promised contributions (PEW Center on Global Climate Change, 2011). Parties also agreed on arrangements for the dissemination and sharing of knowledge and information on adaptation. Principally, the Nairobi Work Programme assumes this role and seeks to improve the understanding and assessment of climate change impacts, vulnerability and adaptation in UNFCCC member countries, along with information on adaptation options, projects and technologies.

In addition to the above funding initiatives under the UNFCCC, some other funds support adaptation activities as well. These include funds such as:

- the Pilot Programme for Climate Resilience administered by the World Bank (operating since 2008) which supports technical assistance and investments to countries seeking to integrate climate risk and resilience into core development planning and implementation;

- the International Climate Initiative of the German Federal Ministry for the Environment, Nature Conservation, Building and Nuclear Safety (operating since 2008) which finances climate and biodiversity projects in developing and newly industrializing countries, as well as in countries in transition;
- the Global Climate Change Alliance (launched in 2007 by the European Commission) to strengthen dialogue and cooperation on climate change between the European Union (EU) and developing countries; and
- the Millennium Development Goals Achievement Fund (MDG-F) (established in 2007 through an agreement between the Government of Spain and the United Nations Development Programme) committed to eradicating poverty and inequality and to accelerating progress towards the Millennium Development Goals (MDGs) worldwide.

The Global Environmental Facility also set up the Strategic Priority on Adaptation Fund, a three-year pilot program to show how adaptation planning and assessment could be practically translated into full-scale programmes and projects. The fund was active from 2004 onwards and is now closed. Furthermore, public and international financial institutions (for example, the KfW Development Bank, the Asian Development Bank and the African Development Bank) commercial banks and insurers have a role to play in financing adaptation. Some commercial banks have begun reviewing their due diligence processes for climate risks and have been engaged in adaptation research over a number of years.

Concluding Comments

This chapter has provided a short timeline of climate policy negotiations, focusing on key policy developments at international levels. Looking at the history of adaptation within the UNFCCC process, it becomes clear that the main intention of the treaty was to focus on mitigation efforts and the development of emission reduction targets. Consequently, the initial focus of climate policy negotiations has rested on mitigation, rather than adapting to the impacts of climate change. A second wave of policy negotiations is now emerging, reflecting a greater attention to adaptation as an important response strategy to climate change alongside mitigation with a primary focus on adaptation in regions most vulnerable to climate change. However, as concerns about adaptation have only just begun to emerge as a more prominent topic on the international climate policy landscape, adaptation objectives have not yet been integrated into key policy targets, leading to uncertainties in the way forward. The following chapter will now look at how these international negotiations have shaped responses to climate change on national and industry levels.

4. Implications for national policy on climate change

The previous chapter provided a short timeline of climate policy negotiations, focusing on key policy developments on international levels. It made the point that the initial focus of climate policy negotiations has rested on mitigation, rather than on adapting to the impacts of climate change. This chapter looks at how international climate policy has been translated into national and sectoral measures. Integrating the dimensions of the Kyoto treaty into national policies and actions on climate change has proven a major challenge to politicians, bureaucrats and regulators. The United Nations Framework Convention on Climate Change (UNFCCC) process has not been without criticism, and the translation of international policy into national outcomes, industry-level applications and local outcomes is a critical step to create incentive structures for organizations and entrepreneurs to mitigate climate change through investments in low-carbon technologies, processes and systems, and to adapt to adverse consequences resulting from climate change. However, are policies designed so that they consider the external and internal forces causing organizations to shift towards action on climate change? The following sections provide an overview over national responses stemming from international climate policy. This chapter highlights how international policy gave rise to the introduction of national measures aimed at mitigation, and then assesses emerging national measures aimed at adaptation. The chapter considers the impact of these measures on organizations and industries.

Mitigation Policies

The Kyoto Protocol asked Parties to implement domestic measures to reduce their greenhouse gas (GHG) emissions. As policies were implemented across Annex B countries, broad 'top-down' approaches such as national emissions trading schemes and carbon taxes (applied to carbon emissions) have been widely adopted. Numerous Annex B countries have opted for the implementation of these 'top-down' approaches to control GHG emissions. Under the Kyoto Protocol, countries are free in their choice of domestic measures. For instance, countries can opt for the implementation of so-called 'sectoral' measures that are applied to specific economic sectors or industries such as energy, agriculture or transportation. Such measures include technical standards, sector-specific limits, or fiscal actions (Montgomery, 2005). Alternatively, countries can also target emission reductions from households, or personal energy use (Pinkse and Kolk, 2009).

Nonetheless, the introduction of a national emissions trading scheme (ETS), targeting the corporate sector and high emitters due to the aggregate emission-intensity of their activities, is often seen as the preferred option. Based on the seminal works of Coase (1960) and Dales (1968), the economic rationale for such a market-based approach is that individual emitters have different marginal costs of emissions control. As such, trading provides an opportunity for emitters to collectively find the least expensive way to achieve emission reduction targets prescribed by governments. These schemes are also argued to promote innovation in encouraging emitters to find less expensive ways of reducing pollution. Consequently, their appeal as a policy instrument has been the intended outcome to minimize the cost of reducing emissions to the economy as a whole (Dargusch and Griffiths, 2008; OECD, 2013).

Putting a price on carbon on the national level: emissions trading
Several countries have opted for the implementation of an ETS. Table 4.1 provides a brief overview of the main schemes in force at the time of writing. While the Kyoto Protocol has established emissions trading between countries as one of the Kyoto mechanisms, it does not automatically require participating countries to also implement a domestic ETS applying to companies (Pinkse and Kolk, 2009).

A domestic ETS works by introducing a limit (or cap) on the aggregate annual emissions from all covered types of emissions and emission-generating activities, which is why these schemes are also referred to as cap-and-trade schemes. This emissions limit can then be lowered over time to achieve a desired national emissions reduction target. Typically, an ETS seeks to capture high-emitting organizations and industries that are issued emission allowances (representing the right to emit a specific amount of GHG emissions). Organizations are then required to surrender a permit for each ton of GHGs they emit. The total amount of available allowances cannot exceed the cap, thus limiting total GHG emissions to the level of the cap. Emissions allowances are tradable, and their price is determined by demand and availability in the market. An ETS is meant to fix the overall emission level of domestic emissions while allowing market forces to set the price of permits. In an ideal world, emissions

Table 4.1   National emissions trading schemes (in force)

Title of Scheme	Australia's Carbon Pricing Mechanism (AUS CPM) (repealed)	Kazakhstan Emissions Trading Scheme (KAZ ETS)	California Cap-and-Trade Program	EU Emissions Trading System (EU ETS)	New Zealand Emissions Trading Scheme (NZ ETS)	Québec Cap-and-Trade System	Regional Greenhouse Gas Initiative (RGGI)	Shenzhen Pilot System	Swiss ETS	Tokyo Cap-and-Trade Program
Type of ETS	Mandatory with voluntary opt-in	Mandatory with voluntary opt-in	Mandatory	Mandatory	Mandatory with voluntary opt-in	Mandatory	Mandatory	Mandatory	Mandatory with voluntary opt-in	Mandatory
Summary	A three-year fixed price period started in 2012; transition into a fully flexible ETS in July 2015. The Liberal National Party coalition, after winning the election in 2013, moved to repeal the CPM and to establish a Direct Action Plan instead.	A two-year pilot phase started in January 2013. A 2nd phase is planned from 2014–15. Kazakhstan ratified the Kyoto Protocol in 2009, but does not have legal basis for participation in the flexible Kyoto mechanisms, such as CDM and JI.	First compliance period from 2013–14. Covers sources responsible for 85% of California's GHG emissions. California aims to reduce GHG emissions to 1990 levels by 2020 and achieve an 80% reduction from 1990 levels by 2050.	Launched in 2005. Covers emissions from 27 EU Member States, plus Croatia, Iceland, Norway and Liechtenstein. The EU ETS underwent major changes as Phase III started in 2013. It experienced a price drop on the carbon market.	The New Zealand emissions trading scheme (NZ ETS) was launched in 2008. The first sector was forestry, followed by liquid fossil fuels, stationary energy and industrial process in 2010, and waste and synthetic GHGs in 2013.	Introduced in 2012 (with a transition year without mandatory compliance). Québec is a member of the Western Climate Initiative (WCI) and will formally link its system with California's in 2014.	First compliance period from 2009–11. Second compliance period from 2012–14. Covers Connecticut, Delaware, Maine, Maryland, Massachusetts, New Hampshire, New York, Rhode Island and Vermont.	The Shenzhen ETS started officially in June 2013 as the first of the Chinese pilot ETS, covering 635 medium- and small-sized emitters from 26 sectors and 200 buildings, accounting for about 38% of Shenzhen's 2010 emissions.	Five-year voluntary phase from 2008 as alternative to CO_2 levy on fossil fuels. Revised ahead of 2013–20 mandatory phase (for large, energy-intensive firms). Medium-sized firms can also opt-in. Participants are exempt from the CO_2 levy.	Launched in April 2010 across Tokyo Metropolis as part of the city's Climate Change Strategy. Regulates industrial/ commercial GHG emissions of large buildings. First compliance period is FY2010–14.

Title of Scheme	Australia's Carbon Pricing Mechanism (AUS CPM) (repealed)	Kazakhstan Emissions Trading Scheme (KAZ ETS)	California Cap-and-Trade Program	EU Emissions Trading System (EU ETS)	New Zealand Emissions Trading Scheme (NZ ETS)	Québec Cap-and-Trade System	Regional Greenhouse Gas Initiative (RGGI)	Shenzhen Pilot System	Swiss ETS	Tokyo Cap-and-Trade Program
Country/region emissions[1]	552 MtCO$_2$e (2011)	212 MtCO$_2$e (2011)	451 MtCO$_2$e (2010)	4550 MtCO$_2$e (2011)	73 MtCO$_2$e (2011)	82 MtCO$_2$e (2009)	679 MtCO$_2$e (2010)	83 MtCO$_2$e (2010)	50 MtCO$_2$e (2011)	62 MtCO$_2$e (FY 2010)
% of which covered	60%	77%	40%	45%	50%	29%	22%	38%	11%	18%
Overall GHG reduction target	By 2012: 108% of 1990 GHG levels (Kyoto target). By 2020: Unconditional target: −5% GHG levels; conditional target: −15% to −25% of 2000 GHG levels. By 2050: −80% below 2000 levels.	By 2020: −5% below 1990 GHG levels.	By 2020: 1990 GHG levels. By 2050: −80% of 1990 GHG levels.	By 2012: −8% below 1990 GHG levels (EU-15 Kyoto target). By 2020: −20% below 1990 GHG levels (EU-27, by European law). By 2050: −80/−95% below 1990 GHG levels (not binding).	By 2012: Stabilization at 1990 GHG levels (Kyoto Protocol). By 2020: −5% below 1990 GHG levels (unconditional target). By 2050: −50% below 1990 GHG levels.	By 2020: −20% below 1990 GHG levels (as part of the 2013–20 Climate Change Action Plan).	RGGI members do not share a common overall GHG reduction target.	By 2015 (Guangdong Province 12th, Five-Year Plan): 21% reduction in carbon intensity compared to 2010.	2012: Switzerland's Kyoto target is −8% below 1990 GHG levels. 2020: At least −20% below 1990 GHG levels (unconditional). Commitments of up to by 40% depend on international agreements.	By 2020: −25% target

Cap and trajectory	Absolute Cap. In 2014, the Government will set caps for five years and then each year will extend the caps by one year (that is, caps for five years will always be set). However, the CPM was repealed.	Absolute Cap. An absolute cap of approx. 147MtCO2 (+ reserve of 20.6MtCO2) was set for 2013. This cap will decrease in a linear fashion in the second phase to achieve the 2020 target.	Absolute Cap. Annual allowance budgets (in Mio. MtCO2e) for the First period: 162.8 (2013), 159.7 (2014), 394.5 (2015, increased scope), decreases by 3% annually to reach a cap of 334.2MtCO2e in 2020.	Absolute Cap. Centralized EU-wide cap for stationary sources (since 2013/Phase III): 2,084MtCO2e, reduced by 1.74% annually (also beyond 2020). Aviation sector cap: 210MtCO2e/year for 2013–20.	Soft cap linked to Kyoto target (61.9Mt CO2e/year excluding LULUCF).	Absolute Cap. First compliance period: 2013: 23.2MtCO2e, 2014: 23.2MtCO2e; Second compliance period (2015–17): 2015: 65.3MtCO2e; decreasing by 2.1Mt CO2e annually.	Absolute Cap. Stabilization at 149.7MtCO2 (2009–14); reduced annually by 2.5% (from 2015); total reduction through 2018: 10%.	Absolute Cap.	Absolute Cap. Voluntary phase (2008-2012): Each participant got its own entity specific reduction target. Mandatory phase (2013-2020): Cap annually reduced by 1.74% based on its 2010 level.
									Absolute cap set at facility-level that adds up to Tokyo-wide cap.
GHG covered	CO_2, CH_4, N_2O, PFCs from aluminum production.	CO_2	CO_2, CH_4, N_2O, SF_6, HFC, PFC, NO_3 and other fluorinated GHG.	CO_2, CH_4, N_2O, SF_6, HFC, PFC	Phase I (2005–07): CO_2; Phase II (2008–12): CO_2 + opt-in of N_2O: Phase III (2013–20): CO_2, N_2O, PFCs.	CO2, CH4, N_2O, SF6, HFC, PFC, Nitrogen trifluoride (NF3)	CO_2	CO_2, and since 2012 N_2O as well as theoretically PFCs (but no production of primary aluminum in Switzerland to date)	CO_2 emitted by the use of fuels, heat and electricity (excluding those for residential purposes)

Title of Scheme	Australia's Carbon Pricing Mechanism (AUS CPM) (repealed)	Kazakhstan Emissions Trading Scheme (KAZ ETS)	California Cap-and-Trade Program	EU Emissions Trading System (EU ETS)	New Zealand Emissions Trading Scheme (NZ ETS)	Québec Cap-and-Trade System	Regional Greenhouse Gas Initiative (RGGI)	Shenzhen Pilot System	Swiss ETS	Tokyo Cap-and-Trade Program
Sectors covered	Electricity generation (fuel combustion), other stationary energy combustion, fugitive emissions (oil, gas, coal), industrial process, waste and wastewater. General emissions threshold: >25,000tCO$_2$ e/year of Scope 1 (direct) emissions.	Energy sector (incl. oil and gas,) mining and chemical industry (>20,000tCO$_2$/year. Possibility of voluntary opt-in of additional sectors.	First compliance period (2013–14): Electric utilities and large industrial facilities. Second compliance period (2015–17): Sectors from first compliance period + transportation, natural gas and other fuels.	Power/heat generation (>20MW thermal capacity), energy-intensive installations, commercial aviation from 2012 (>10,000 tCO$_2$/year; exemptions for flights to/from non-ETS countries). Since Phase III: CCS installations + new sectors (incl. aluminum).	Forestry (since 2008), stationary energy (since 2010), industrial processing (since 2010), waste (since 2013; except small and remote landfills), synthetic GHG (since 2013). Agriculture currently has a reporting obligation.	2013–14: Electricity, Industry (>25,000tCO$_2$e/ year); 2015–17 and 2018–20) First compliance period sectors + distribution and import of fuels for consumption in the transport and building sectors, as well as in small and medium-sized businesses, threshold: >25,000tCO$_2$e/ year.	Fossil Fuel Electric Generating Units (Threshold: >25MW).	26 sectors, including electricity generators, industrial companies and building sector. Inclusion threshold: 20,000tCO$_2$/ year considering both direct and indirect emissions. Inclusion of transport sector under consideration.	Mandatory participation: Industries listed under Annex 6 of the revised CO$_2$ Ordinance (25 subsectors). They generally have a total rated thermal input of >20MW. Possible voluntary opt-in for certain sectors.	Commercial sector: Office buildings, public buildings, commercial buildings, heat suppliers Industrial sector: Factories, sewage and waste management General threshold: Facilities that consume more than 1,500 kilolitres of crude oil equivalent or more.

Number of liable entities	As of September 2013, 375 entities.	Approx. 178 businesses.	Approx. 350 entities, when the program is fully operational.	About 12,000.	As of June 2013, 187 mandatory +2,580 voluntary participants.	Approx. 80 operators for 2013–2014.	Currently 168 entities covered.	635 entities.	About 40–50.	1,348 entities for FY2010.
Allocation	A significant proportion of units will be auctioned starting from FY 2013/14.	100% free allocation in pilot, based on 2010 emissions.	Allowances are allocated by benchmarks in each sector, remainder is auctioned.	>40% of allowances auctioned in 2013 (100% for electricity sector). Free allocation, based on EU-wide benchmarks, historical activity data, a carbon leakage exposure and a reduction factor.	Intensity based allocation: 90% for high emissions-intensive activities; 60% for moderate emissions-intensive activities.	2013–14: Free allocation based on historical levels, production levels and intensity target. Diminishes by about 1–2% over 2015–17. 100% auctioning for electricity and fuel distributors.	Approx. 90% of all allowances offered at auction (using a 'single-round', 'sealed-bid', 'uniform-price' format). Rest of allowances held as reserve.	Allowances are distributed for free, based on a game theory approach that takes into account the companies' own estimations of output and emissions. Auctioning in future.	Mandatory phase (2013–20): Free allocation is based on industry benchmarks using a similar methodology as in the EU ETS. No free allocation for the power sector.	In general, grandfathering based on historic emissions; new entrants: based on past emissions or based on emission intensity standards.

Title of Scheme	Australia's Carbon Pricing Mechanism (AUS CPM) (repealed)	Kazakhstan Emissions Trading Scheme (KAZ ETS)	California Cap-and-Trade Program	EU Emissions Trading System (EU ETS)	New Zealand Emissions Trading Scheme (NZ ETS)	Québec Cap-and-Trade System	Regional Greenhouse Gas Initiative (RGGI)	Shenzhen Pilot System	Swiss ETS	Tokyo Cap-and-Trade Program
Assistance to industry	Transitional assistance for Emissions Intensive Trade Exposed firms and coal-fired generators.	Not known.	Transitional assistance for industrial facilities in form of free allowances.	Phase 1: 95% of permits allocated for free; Phase 2: 90% allocated for free; Phase 3: progressive shift towards auctioning of allowances in place of cost-free allocation.	Some allowances allocated for free for trade-exposed industrial sectors, forestry and fishing.	Several sectors subject to international competition will receive some free allowances.	None.	Free allocation.	Similar to EU ETS.	Free allocation.

Monitoring, Reporting, Verification (MRV)	Mandatory reporting of GHG emissions, energy production and consumption for facilities above certain thresholds. Legislated under the National Greenhouse and Energy Reporting Act 2007. The Clean Energy Regulator may conduct audits to verify data.	Mandatory reporting for entities above 20,000tCO$_2$/year. Reporting also required for CH$_4$ and N$_2$O. Verification by accredited third-party verifiers. Installations below the threshold must submit inventory reports (non-verified).	Mandatory reporting for most entities above 10,000tCO$_2$. Verification by independent third-party verifiers annually or triennially.	Monitoring Plan required for every installation. Annual self-reporting based on harmonized electronic templates prepared by the European Commission. Verification by independent accredited auditors required before 31 March of each year.	Annual self-reporting. Verification by a third party is required only when participants apply for use of a unique emission factor.	Reporting is required above 10,000tCO$_2$. Verification of emissions above 25,000tCO$_2$ via an organization accredited to ISO14065.	Emitters report quarterly CO$_2$. Data are transferred to the RGGI CO$_2$ Allowance Tracking System (RGGI COATS, available for public view).	Monitoring, Reporting and third-party Verification are regulated by the Shenzhen GHG Monitoring and Reporting Regulation and Reporting Guideline and the Shenzhen GHG Verification Regulation and Guideline.	Annual self-reporting on emissions. Monitoring Plans are required for every installation. The Federal Office for the Environment may order verification of the reports Monitoring Plans by a third party.	Participants are required to report annually (fiscal year) their verified emissions.

Title of Scheme	Australia's Carbon Pricing Mechanism (AUS CPM) (repealed)	Kazakhstan Emissions Trading Scheme (KAZ ETS)	California Cap-and-Trade Program	EU Emissions Trading System (EU ETS)	New Zealand Emissions Trading Scheme (NZ ETS)	Québec Cap-and-Trade System	Regional Greenhouse Gas Initiative (RGGI)	Shenzhen Pilot System	Swiss ETS	Tokyo Cap-and-Trade Program
Offsets	Domestic offsets through Carbon Farming Initiative.	Two types of domestic offsets are allowed.	Certain domestic offset types are accepted.	–	–	Three domestic offset types are accepted: manure methane, landfill methane, ozone depleting substance removal/destruction from refrigerators.	Five domestic offset categories: landfill methane, sulphur hexafluoride (SF6) reduction, forestry, energy efficiency, manure management.	Project-based government approved carbon offset credits.	–	Credits from four offsets types are allowed.

Credits	International credits (CERs, ERUs) can be used in the flexible price period, quantitative restrictions apply.	International credits may be allowed in the future.	–	CDM/JI credits. Quantitative and qualitative restrictions apply.	ERUs, RMUs and CERs are allowed, but qualitative restrictions apply similar to the EU ETS.	–	–	–	ERUs and CERs are allowed. Qualitative restrictions apply including those in the EU ETS. The use of ERUs and CERs is also subject to quantitative restrictions.	–
Linkages with other Schemes	A one-way link between the Australian ETS and the EU ETS was planned from 2015 onwards, and a full two-way link from 2018 onwards.	–	California will formally link its system with Québec's in 2014 as part of the Western Climate Initiative (WCI).	Possible link with Australian ETS. The Commission is negotiating with Switzerland on linking the EU ETS with the Swiss ETS.	In December 2011, New Zealand and Australia announced that they were exploring linking their schemes.	Québec will formally link its system with California's in 2014 as part of the Western Climate Initiative (WCI).	–	–	Possible link to the EU ETS. Many elements of the Swiss ETS have been designed following the EU ETS (for example, allocation benchmarks).	Linking with the Saitama Prefecture since its launch in April 2011. Trade of credits from excess emission reductions to commence in 2015.

Note: [1] 2013 UNFCCC data except for regions.

Source: Adapted from the International Carbon Action Partnership (2014).

trading would occur globally and establish a uniform carbon price for all countries and sectors, such that emissions can be reduced where it is cheapest to do so. Currently, however, no global emissions trading scheme is in sight, except for some discussions around possible linkages between emissions trading schemes.

The introduction of cap-and-trade systems has led to much discussion around the instrument's environmental effectiveness, economic costs and distributional impacts. There is considerable debate around the 'best' ETS design in terms of cap-setting, scope and coverage of GHG emissions and sectors, allocation decisions (and industry assistance), monitoring, verification and enforcement, as well as options such as credits and offsets that expand mitigation options beyond the region and/or sectors covered by an ETS. Scientific studies have primarily focused on the experience with Phases I and II of the EU ETS (2005–07 and 2008–12) as the EU ETS is the largest scheme in operation. The main pillars of the EU ETS were built on years of economic research into theories of emissions trading (and some practical experience with emissions trading schemes in the US); however, the practical experiences with an ETS unprecedented in scale and scope highlighted many profound differences between theory and practice (Grubb and Neuhoff, 2006).

For instance, experience with the EU ETS has shown the importance of allocating allowances. As evident from Table 4.1, most schemes provide some assistance to industry, often in the form of free allowances, in order to facilitate the transition to an ETS or to support trade-exposed, emissions-intensive industries. Phase I of the EU ETS was characterized by an over-allocation of emission permits, which resulted in a rapid collapse of the EU allowance price after first data on verified emissions were published. This was attributed to the fact that most allowances were allocated for free based on historical emissions, which is also known as 'grandfathering'. As a result, the European Commission tightened the regulations of the EU ETS for Phase II (2008–12) (Brunner et al., 2011). While the EU ETS enjoyed two years of relatively low volatility and relatively stables prices from 2009–11, the ongoing economic crisis, along with renewed issues due to oversupply and an increase in international credits, led to a sharp decline from June 2011 onwards (World Bank, 2013a). At the start of Phase III in 2013, the surplus stood at almost two billion allowances, eventually leading to a crash of the carbon price in 2013. A legislative proposal to improve market stability was put forward in January 2014, together with a proposal for a 2030 framework for climate and energy policy. Both proposals were still requiring approval by the Council and the European Parliament to become law at the time of writing (European Commission, 2014).

Putting a price on carbon on the national level: carbon tax

The alternative to the implementation of an ETS is a carbon tax, defined as a tax using a metric directly based on carbon (namely, a tax per tCO_2e or tons of carbon dioxide equivalent) (World Bank, 2013a). In contrast to an ETS, a carbon tax fixes the price while the emission level is allowed to vary according to economic activity. Countries can choose to implement an ETS as the only instrument to put a price on carbon or use taxes as the sole instrument, or use taxes in conjunction with an ETS. European countries with carbon taxes use this measure in a complementary way to the existing EU ETS. Several countries and regions apply carbon taxes, including Australia, British Columbia, Denmark, Finland, Ireland, Japan, Norway, South Africa, Sweden, Switzerland and the United Kingdom. In Scandinavia, carbon taxes have been in place since the 1990s, while other countries have introduced taxation more recently. The taxes can be either implemented economy-wide, or target/exclude certain sectors (sectoral measures).

Examples of carbon taxes implemented around the world are as follows (World Bank, 2013a):

- In Australia, the Carbon Pricing Mechanism is a three-year fixed price period prior to the implementation of an ETS. (This was repealed in July 2014.)
- In British Columbia, the Revenue Neutral Carbon Tax was introduced in 2008, targeting all consumers of fossil fuels.
- In Denmark, fossil fuels are subject to both an energy tax and a CO_2 tax that commenced in 1992. The tax differs according to the type of fuel. It applies to households and industries, but with partial exemptions for operators covered under the EU ETS as per EU Energy Tax Directive and energy-intensive industries that enter voluntary agreements on energy efficiencies.
- In Finland, a CO_2 tax was introduced in 1990 and applies to all consumers of fossil fuels. It differs by the type of fuel and its carbon content, with exemptions for certain industries and fuel types.
- In the Republic of Ireland, the Natural Gas Carbon Tax (starting 2010), the Mineral Oil Tax (starting 2010) and the Solid Fuel Carbon Tax (starting 2013) apply to all consumers of fossil fuels, with partial exemptions for operators covered under the EU ETS as per EU Energy Tax Directive.
- In Japan, the Tax for Climate Change Mitigation was introduced in 2012 and targets all consumers of fossil fuels (with exemptions for certain parts of the agriculture, transport and industry sectors).

● In Norway, the CO_2 tax was introduced in 1991 and targets all consumers of mineral oil, gasoline and natural gas. Both the EU ETS and CO_2 tax are imposed on offshore production and distribution of oil and gas, but other operations not in this sector and covered by the EU ETS as well as certain other industries are (partially) exempt.

Effectiveness of Policy Approaches

The Kyoto Protocol has generally been seen as a significant first step towards a global regime to stabilize and reduce GHG emissions. Governments put legislation and policies in place to meet their emission reduction commitments. This has led to the creation of carbon markets (for example, the EU ETS) and business-related investment decisions. However, the different policy measures have led to significant debates in regards to their effectiveness in achieving emission reductions. The biggest points commonly raised are:

1. **Does the Kyoto Protocol bring about enough momentum for change on a organizational and industry level?** The Kyoto Protocol is largely based on neoclassical economic thinking and leaves it up to individual countries to implement domestic GHG emissions reduction initiatives. Critics have argued that the Kyoto Protocol has not created a strong and direct nexus between mitigation policies and organizational- and industry-level innovation (based on evolutionary economic theory). There also is not a strong and direct nexus between global emission targets and organizational-level mitigation targets. The argument is that the Kyoto Protocol has not only failed to provide strong mitigation targets, but has also not been a strong impetus for rapid industry transformations to a new, low carbon economy. Gilding (2011), for instance, states that: 'It is not simply about fossil fuels and carbon footprints. We have come to the end of economic growth, Version 1.0, a world economy based on consumption and waste'.

    Importantly, carbon policies are not the only instrument influencing mitigation outcomes. Other important factors are direct links between carbon emissions and the level of economic growth, as well as national targets and policies driving the uptake of renewable energy (for example, feed-in tariffs, industry regulation and subsidies). A report by the European Commission (2012) put forward that:

The ETS will be critical in driving investments in a wide range of low carbon technologies. It is designed to be technology neutral, cost-effective and fully compatible with the internal energy market. The ETS will need to play an increased role in the transition to a low-carbon economy by 2050. Since the start of the second trading period in 2008, emissions are down by more than 10 per cent but while the carbon price signal of the EU ETS has certainly contributed to this, the economic crisis is clearly the major cause of these strong emission reductions.

The validity of this claim was questioned by those who argued that Europe's energy policy, and not the economic slowdown, was the most influential factor in the reduction of carbon emissions.

According to an analysis by French investment group CDC Climat (owned by France's sovereign wealth fund the Caisse des Dépôts), energy policies in Europe led to a major uptake of renewables that were more significant in reducing carbon emissions explain half of the carbon savings since 2005, representing 500 million tons of avoided CO_2 emissions. The EU adopted new energy policies in 2009 which included a series of measures to reduce CO_2 emissions by 20 per cent by 2020, while increasing energy efficiency by 20 per cent and the share of renewable energy to 20 per cent (the co-called 20/20/20 target). CDC Climat argues that the economic slowdown contributed only to a smaller share of emission reductions, estimated at 300 million tons of avoided CO_2 emissions. At the same time, the energy policies in Europe may have come into conflict with climate policy, as the high uptake of renewables may have ultimately undermined the carbon market.

Writing in the scientific journal *Nature*, Prins and Rayner (2007) argued that the Kyoto Treaty has become a limiting factor in addressing the issue of emission reductions, and may not altogether be the most suitable mechanism. Predictably, conservative commentators have used this view to articulate a case that the Kyoto Protocol does not work and should be abandoned. However, these arguments whitewash the nuances of the debate and the perspective originally put forward by Prins and Rayner. Their argument is that:

Kyoto's supporters often blame non-signatory governments, especially the United States and Australia, for its woes. But the Kyoto Protocol was always the wrong tool for the nature of the job. Kyoto was constructed by quickly borrowing from past treaty regimes dealing with stratospheric ozone depletion, acid rain from sulphur emissions and nuclear weapons. … Unfortunately, this borrowing simply failed to accommodate the complexity of the climate-change issue.

The argument by Prins and Rayner is clear – we need to address the problems of GHG emissions. However, grand targets and national goals will only work in the context of developing and encouraging 'social learning' and experimentation at local and regional levels – so that a focus on what governments, communities and organizations actually do is what becomes important.

2.   **Are the Kyoto targets ambitious enough?** The targets under the first commitment period under Kyoto added up to an average 5 per cent of emission reductions compared to 1990 levels over the five-year period 2008 to 2012.[1] This compares to estimates that a 450ppm objective (often regarded as synonymous with limiting mean global temperature by 2100 to no more than 2°C above pre-industrial levels) would require developed countries to commit to emission reductions of 32 per cent by 2020 over Kyoto levels (2008–12), or around 5 per cent reductions per year (Garnaut, 2011). Some of the largest emitters were and are still not captured under the Kyoto Protocol. For example, the United States as one of the largest emitters of carbon dioxide emissions (the largest emitter until 2006, when China overtook it) is a non-signatory. The Doha Amendment to the Kyoto Protocol provides for the continuation of legally binding GHG emission reduction targets, the second-round targets encompass barely 15 per cent of global emissions.

As the costs of emission reductions are unknown, and possibly large, countries with substantial emissions had a vested interest to make sure that targets were lax. McKibbin and Wilcoxen (2009) attributed the US Senate's lack of support for ratifying the Kyoto Protocol (see previous chapter) to a critical flaw in its design given that it requires participating industrialized countries to agree to achieve a specified emissions target regardless of the costs of doing so. Countries with significant emissions profiles attempted to negotiate more favorable targets as a condition for their continued participation. Japan, Canada and Russia, for example, were able to negotiate great increases in their 'sink' allowances during COP-6 and COP-7. Consequently, the current international policy context on climate change can hardly be described as a level playing field (Pinkse and Kolk, 2009). The lack of more stringent targets and the ongoing growth of overall emissions on a global scale have led to criticisms that Kyoto has ultimately not produced demonstrable

[1]   At the time of writing, the Doha Amendment to the Kyoto Protocol has not yet entered into force.

reductions in global GHG emissions or in the anticipated emissions growth (Prins and Rayner, 2007).

3. **What is the most effective policy approach?** There have also been significant discussions among policy-makers and business leaders on the impacts of mitigation policies on the economy and different industry sectors on competitiveness, corporate perform-ance and technological change towards a low-carbon economy. The implementation of different policy approaches (carbon taxation versus emissions trading) has prompted debates in regards to the most efficient response mechanism for GHG emission reduction efforts on an economic level, with economic arguments favoring the introduction of an ETS over carbon taxation (OECD, 2013). Within some national contexts, the climate policy debates have focused entirely on the costs for carbon-intensive industries resulting from the introduction of legislation, rather than on the opportunities associated with investing in low-carbon sectors (Foxon and Andersen, 2009). In Australia, for instance, most of the current debate surrounding the implementation on climate policies has focused on who should bear the economic burden, rather than on tackling climate change.

Irrespective of criticism of the UNFCCC process and the Kyoto Protocol, it provides nonetheless an important international archi-tecture for countries to engage in negotiations – which will most likely gain in importance as the impacts of climate change become more visible. In order to prevent the adverse effects of climate change, both industrialized and developing countries will need to agree to deep emission cuts, or significantly strengthen efforts at adaptation, as discussed below. Looking at the historic responsibil-ity and economic capabilities of industrialized countries, an argu-ment can certainly be made that these countries should take a lead in mitigation. However, the engagement of developing countries will become increasingly important, especially of those countries with significant emission profiles. The UNFCCC also acknow-ledges the need not just to reduce emissions, but also to protect forests as part of the efforts to combat climate change. Tropical deforestation was initially excluded from the Kyoto Protocol due to controversies surrounding sovereignty and scientific uncertainty, but discussions on reducing emissions from deforestation in developing countries are now well underway within the UNFCCC process, at the initiative of developing countries (UNFCCC, 2009).

Any significant progress on climate change can only be resolved through aggregate actions across a variety of different actors. This will require forging agreements between governments, corporations and communities on the progress that is to be achieved (Dunphy et al., 2003). According to the IPCC, there is significant mitigation potential through energy efficiency and the uptake of renewables for all sectors. Mitigation costs in 2030 would not exceed 3 per cent of global GDP and could yield multiple social and environmental benefits (UNFCCC, 2009). However, the complex issue of action on climate change will have to be addressed through a mix of public and private actors, including national governments, corporations, multinational organizational, industry associations as well as empowered public interest groups. This kind of concerted action will be needed in addition to intergovernmental negotiations on how to tackle climate change, particularly because the primary target of these negotiations is not directly aimed at the level of individual actors. It is therefore not surprising that some deficits have emerged as a result of 'top down' regulation such as carbon taxes and market-based measures such as tradable permits.

4. **Can underlying disagreements over equity and burden sharing between countries be resolved?** The UNFCCC and the Kyoto Protocol have been based on the fundamental idea that all of the world's governments need to be engaged in tackling climate change, thus matching a global threat with a universal global response. However, the large number of countries involved has lowered the common denominator for agreement (Prins and Rayner, 2007), and the focus on finding a policy solution has shifted attention away from the organizational and industry levels where innovation and sustainable solutions can be implemented. One major area of disagreement has been over equity and burden sharing – particularly as countries' welfare differs significantly, and as not all countries have equally contributed to (and are equally affected by) climate change. While the economic development and progress in developed countries have been key drivers of climate change, several commentators have raised concerns that international treaties such as Kyoto are not effective if they do not tackle Parties with significant emissions growth, such as developing and newly industrialized countries with significant population growth and rapid economic progress.

Much of the future growth in emissions has not been captured under Kyoto. At the time when climate change negotiations culminated in the UNFCCC, more than 50 countries that were classified as 'developing' (Non–Annex I) had higher per capita incomes on a purchasing power parity basis than the lowest income 'developed' (Annex I) country. In 1995, by the time of the Berlin Mandate, Non–Annex I Parties had higher total annual GHG emissions than Annex I Parties. And in 2005, by time the Kyoto Protocol entered into force, nearly 50 Non–Annex I Parties had per capita fossil-fuel CO_2 emissions that exceeded those of the Annex I countries with the lowest per capita measures (Aldy and Stavins, 2012). Some estimates suggest that, in the first decade of the present century, China emitted more CO_2 from fossil-fuel combustion than any developed country had emitted over the course of the twentieth century – with the two exceptions of the United States and Germany (Aldy and Stavins, 2012). If and how these issues can be resolved will have to been seen in future climate change negotiations – which are trying to achieve a global outcome.

5. **Integration of adaptation and mitigation:** Few attempts have been made to ensure that policies intended to mitigate climate change also support adaptation and mitigation objectives. This comment does not just apply to international policy negotiations, but also to strategies on organizational, industry and national levels. At organizational and industry levels, companies have mostly focused on responding to the impact of climate policy and mitigation requirements through either *innovation strategies* or *compensation strategies* (Pinkse and Kolk, 2009). However, these strategies for mitigating emissions do often not include particular considerations regarding an organization's or sector's company's physical vulnerability to climate change, and thus neglect longer-term strategies focusing on adaptation and resilience.

    Innovation strategies can include energy- or GHG savings measures, for example, reduced air-conditioning, heating or lighting, or reduced use of PFCs in industrial processes. They can also include the implementation of new technologies such as solar power, cogeneration or combined heat and power (CHP). Other options are optimization attempts along supply chains, including life-cycle approaches (that is, reducing climate impacts associated with all the stages of a product's life from-cradle-to-grave, including raw material extraction, manufacturing, distribution, use, maintenance, and disposal or recycling). Lastly, partnerships or consortia offer new product/market combinations. One example is the California Fuel Cell Partnership, a collaboration of auto manufacturers, energy providers, government agencies and fuel cell technology companies that work together to promote the commercialization of hydrogen fuel cell vehicles (Pinkse and Kolk, 2009). Compensation strategies include engagement in carbon markets or the external acquisition of lower-emissions technologies. Larger companies operating across borders can internally transfer emissions or set up internal carbon trading mechanisms (divisions inside a company trade with each other to meet their mitigation targets so that reductions can occur wherever they are cheapest). Compensation can also occur by outsourcing or subcontracting emissions-intensive activities, thus shifting the burden to other companies (Pinkse and Kolk, 2009).

The challenge will be to develop policy approaches that support adaptation, mitigation and resilience at the organizational and industry level. Efforts to mitigate GHG emissions over the next few decades can substantially reduce the risks of climate change in the second half of the twenty-first century (IPCC, 2014c). At the organizational and industry level, mitigation strategies in the form of innovation and compensation are often variations of established competitive corporate responses (such as process improvements, efficiency gains or outsourcing) and bring outcomes in the short- and medium-term due to costs-savings, the avoidance of fines (by complying with carbon regulation) and innovation. However, mitigation strategies alone do not address impacts resulting from the warming that would occur even for the lowest stabilization scenarios assessed (Pachauri and Reisinger, 2007). Strategies aimed at organizational and industry adaptation and resilience to climate change impacts provide mechanisms to protect against future risk – and require evaluations about costs and future benefits. Adaptation and resilience mechanisms require upfront investments (and carry risks associated with inaccurate prediction), while benefits become only visible in the long-term as the impacts of climate change become more noticeable. The benefits of mitigation and adaptation thus occur over different timeframes.

Mitigation measures are much easier to implement as they are more clearly defined. The literature on mitigation costs and benefits is much more comprehensive, and there are clear metrics for assessing the effectiveness of mitigation measures, such as the reduction in GHG emissions per unit of output. In contrast, what does and does not constitute adaptation is much more ambiguous and dependent on the local context. The literature on adaptation costs and benefits is much less developed and contested, and there are no widely accepted metrics or benchmarks for assessing the effectiveness of adaptation actions, policies and measures. At the same time, adaptation decisions are largely decentralized, and will affect corporate investment or local government planning. As a result, they are more difficult to subsume under national and international coordination compared to mitigation efforts, also because a focus on adaptation would mean to acknowledge that failed efforts at mitigation bring about a greater need for adaptation internationally (Shardul and Samuel, 2008).

Following on from this discussion, the subsequent sections examine the national-level adaptation arrangements in several countries, focusing on overarching policy approaches or frameworks that provide information, guidance or statutory frameworks to facilitate adaptation across actors. Actors can include, for example, individuals, businesses and communities, but also government entities such as those responsible for infrastructure decisions, or the delivery of government services (Productivity Commission, 2012). As becomes evident from these sections, most current adaptation plans only provide high-level guidance to governments, communities, organizations and other actors, without putting any stringent adaptation guidelines into place. Nonetheless, some local approaches and community-level adaptation plans are emerging, mostly in response to local vulnerabilities. While some level of adaptation is taking place, stringent efforts across organizations, industry and society are not yet evident in many instances.

Adaptation Plans in Developing Countries

The previous sections have shown that mitigation has received much more attention than adaptation within the UNFCCC, and that the UNFCCC process has primarily focused on supporting adaptation in the developing world. The UNFCCC process has supported in particular the least developed countries in preparing and implementing National Adaptation Programmes of Action (NAPAs) with a focus on urgent and immediate needs, and on developing National Adaptation Plans (building on existing NAPAs) with a focus on identifying medium- and long-term adaptation needs. At the time of writing, the UNFCCC Secretariat had received 50 NAPAs from the least-developed countries. The database of NAPAs is available via the UNFCCC website. Submission of the NAPA document determines eligibility to apply for funding for implementation under the Least Developed Countries Fund, which is managed by the Global Environment Facility (GEF).

National plans for adaptation in Least Developed Countries (LDCs) can be found in the NAPAs. Other developing countries that are not classified as 'least developed' (for example, South Africa) do not produce

NAPAs, but have at times started to develop their own adaptation policies and frameworks. Hardee and Mutunga (2010) conducted an extensive review of 41 NAPAs submitted by LDCs to the UNFCCC. Following NAPA guidelines, countries undertake four steps to develop their NAPAs: first, the establishment of a NAPA organization which should include local consultation; second, a synthesis of available information on the impacts of climate change and coping strategies to provide a baseline measure of vulnerabilities; third, the identification of priority projects through stakeholder consultation; and fourth, the submission of the NAPA to the UNFCCC. The guidelines for the preparation of the NAPAs (UNFCCC, 2002) emphasize the importance of mainstreaming NAPAs once established, and state that:

> mainstreaming refers to the integration of objectives, policies, strategies or measures outlined within a NAPA such that they become part and parcel of national and regional development policies, processes and budgets at all levels and at all stages, and such that they complement or advance the broader objectives of poverty reduction and sustainable development

Hardee and Mutunga (2010) found a total of 448 priority adaptation projects across 41 NAPAs, but with significant variations among the countries. Half of the projects fall into the categories food security, terrestrial ecosystems and water resources, which can be explained by the importance of agriculture, livestock and fisheries for feeding and sustaining livelihoods for millions of people. The smallest number of priority projects is in the tourism, insurance and energy sectors.

While bringing about important progress in identifying vulnerabilities and adaptation needs, the NAPA process has been criticized for a number of reasons. Osman-Elasha and Downing (2007) analyzed 14 NAPAs and found major weaknesses in the NAPA preparation process, including bureaucratic and institutional barriers that hindered information exchange and created communication problems between different levels of government. The NAPAs showed a lack of technical capacity and expertise, as well as inadequate resourcing and finance. Hardee and Mutunga (2010) criticized the low priority given to health and population growth, and concluded that the NAPA process was not successful in aligning NAPAs with urgent and immediate actions to address vulnerability to climate change. The researchers also found a lack of integration of NAPAs with existing development plans on poverty reduction, sustainable strategy, national nature conservation or disaster preparedness. Vincent et al. (2013) concluded that the amount of local consultation is varied in the plans, and that national level policies on adaptation tend to prioritize large-scale infrastructure projects and state-managed technological and natural resource development, while local level adaptation realities (local knowledge, traditional resource management or the potential use of social networks) are often not recognized.

Uganda's NAPA, for instance, details a number of local adaptation strategies and coping strategies, yet the government defines whether these strategies are desirable or undesirable. Such examples highlight the influence of government and state agencies in the NAPA development process, while civil society organizations and communities often only play a limited role in the formulation of adaptation strategies. While the active participation of local communities and other local stakeholders is generally recognized as necessary for successful implementation of any adaptation plans, few thoughts have been given to establishing suitable frameworks to engage meaningfully with these stakeholders, and to incentivize their active involvement (World Resources Institute, n.d.). There are also no significant resources for systematically involving local stakeholders, raising awareness, building capacities and creating an enabling governance structure for the engagement of citizens and communities. This certainly is a crucial drawback in the development of adaptation plans that may result in subsequent problems in terms of implementing and achieving stated objectives (World Resources Institute, n.d.). These points have repeatedly been highlighted as areas for future attention is policy processes.

Adaptation Plans in Developed Countries

The following sections examine some of the adaptation plans by developed countries. The NAPAs and National Adaptation Plans that have been developed as part of the UNFCCC process thereby need to be distinguished from other types of adaptation frameworks to guide adaptation policy or government processes, especially those from developed countries. These range from statutory frameworks (as in the UK) to adaptation reports that have been commissioned to inform policy responses (as in the US and Canada). In 2012, the OECD concluded that 17 OECD countries had adopted national adaptation strategies (OECD, 2012), but with considerable variation in policy instruments being used and in the assignment of responsibilities between national, state and local governments (OECD, 2012). Not all countries have adopted an explicit policy framework for adaptation. For example, the New Zealand Government mainly provides information and guidance to facilitate adaptation by local governments and others (Productivity Commission, 2012). Below, we outline some examples of adaptation plans, focusing on both developing and least developed countries, as well as developed countries.

Developed country plans: United Kingdom

The United Kingdom is among the countries that have been proactive in facilitating adaptation to climate change. Through the enactment of the Climate Change Act 2008, the UK was the first country to implement a statutory framework for facilitating adaptation. The three main adaptation measures required by the Climate Change Act 2008 (in addition to stringent mitigation targets) are a UK-wide climate change risk assessment, a national adaptation program and adaptation reporting powers which enable the Secretary of State to direct 'reporting authorities' to develop climate change adaptation reports. The Act states that the Secretary of State has a duty to present to parliament within three years of the Act coming into force (and every five years thereafter) a report containing an assessment of the risks for the UK of the current and projected impacts of climate change. The adaptation reporting powers under the Climate Change Act 2008 allows the Secretary of State to direct authorities (that is, organizations with functions of a public nature and statutory undertakers) to prepare climate change adaptation reports. Such reports must explain how these organizations are assessing and reacting to various climate change risks and opportunities (DEFRA, 2012). The Climate Change Act 2008 also established the Committee on Climate Change and the Adaptation Sub-Committee. Both bodies advise the Secretary of State on the risk assessments, assess how prepared the UK is for climate change, and promote action to adapt to climate change. The Adaptation Sub-Committee develops yearly progress reports to assess how the UK is preparing for the risks and opportunities associated with climate change.

The first round of reporting took place between October 2010 and December 2011, and nearly 100 reporting authorities presented reports on their adaptation plans. The first risk assessment report was released in January 2012 and focused on five areas, including the natural environment, agriculture, business, health and wellbeing, building and infrastructure. The findings regarding the impact of climate change on each of the above areas are summarized in Table 4.2. Findings from the reports indicated that the energy, electricity and forestry companies have been actively engaging in industry-level research to determine risks that could arise as a result of climate change. Some organizations also indicated that they have taken active measures to raise awareness of climate change, with some gas companies such as Southern Gas Network producing a questionnaire and an 'Adapting to Climate Change Briefing Pack' (DEFRA, 2012). As a response to the risk assessment report, the UK government developed the National Adaptation Program (NAP) in 2013 (DEFRA, 2013). The NAP report contains an annex that discusses the

Table 4.2   Risk assessment under the UK Climate Change Act 2008

Area	Impact of climate change on area
Natural environment	– Some species of animals and plants are likely to be lost to the UK as their 'climate space' moves. – The arrival of invasive non-native species may pose a threat to native species. – Some species of animals and plants may be better able to adapt than others. – The freshwater environment is projected to come under growing pressure. – Drier conditions projected to lead to a decline in habitats that only develop in a cool, wet climate. – Drier soils may lead to a decline in soil quality. – The risk of wildfires is projected to increase.
Agriculture and forestry	– Warmer temperatures may benefit crop growth, if water is not limiting. – Warmer temperatures projected to provide suitable climatic conditions for new crops in the UK. – Less water projected to be available to meet increased demand for crop irrigation. – Agricultural land projected to become more prone to flooding. – Productivity of commercial tree species projected to change significantly. – Increases in drought, pests and diseases have the potential to reduce timber yield and quality.
Business	– Quicker, cheaper shipping routes could link the UK with key global markets. – Warmer temperatures might increase the UK's appeal as a tourist destination. – Fish and shellfish are projected to shift northwards. – Increased flooding would negatively affect businesses. – Hotter summers projected to increase risk of overheating in workplaces.
Health and wellbeing	– Milder winters predicted to result in a major reduction in the risk of cold-related death and illness. – Hotter summers projected to increase risk of heat-related death and illness. – The number of casualties due to flooding and the impact of floods on mental wellbeing projected to increase. – The risk of health problems caused by marine and freshwater pathogens projected to increase. – Health problems caused by air pollution may increase.
Buildings and infrastructure	– Energy demands for heating projected to decrease. – Energy demands for cooling projected to increase. – Anticipated increase in flood risks to buildings and key infrastructure. – Overheating projected to pose an increased risk to building occupants. – The Urban Heat Island Effect could become more pronounced. – Water resources projected to become scarcer. – Sewers projected to fill more frequently, and spill into rivers and the sea. – Damage to road and rail bridges projected to increase. – Electricity network capacity losses projected to increase.

Source: DEFRA (2012).

role of society in adaptation, the costs and benefits of climate change and the difficulties associated with uncertainty. The NAP puts forward a number of actions related to raising awareness of the need for climate change adaptation, increasing resilience to current climate extremes, taking timely action for long-lead time measures, and addressing major evidence gaps (DEFRA, 2013).

United States

The US Global Change Research Program (USGCRP) is required by the US Global Change Research Act 1990 to 'assist the Nation and the world to understand, assess, predict, and respond to human-induced and natural processes of global change' (US Global Change Research Program, 2013). The USGCRP coordinates federal research across 13 government agencies on changes in the global environment and the consequences of such changes on society. The USGCRP aims to meet four goals through its coordination of research (US Global Change Research Program, 2013): first, to advance science; second, to inform decisions; third, to conduct sustained assessments; fourth, to communicate and educate.

Since 1989, USGCRP has produced an annual report called *Our Changing Planet*. The report is also required by the Global Change Research Act 1990 and outlines recent programmatic achievements, near-term plans and progress in implementing long-term goals. US states and municipalities are recognizing the importance of undertaking pre-emptive measures to address individual vulnerabilities to the impacts of climate change, and have begun to address adaptation needs through climate adaptation plans (Center for Climate and Energy Solutions, 2014). As of December 2013, 15 states in the US have adaptation plans in place, and five states are in the progress of developing adaptation plans (Center for Climate and Energy Solutions, 2014).

European Union

The European Union tended to primarily focus on delivering on the Kyoto targets and initially gave rather limited attention to the role of adaptation. However, as climate change impacts became more visible, the European Commission acknowledged the need for comprehensive adaptation strategies in Member States. In 2007, the European Commission released its Green Paper *Adapting to Climate Change in Europe – Options for EU Action*. In 2009, the European Commission released a subsequent White Paper titled *Adapting to Climate Change: Towards a European Framework for Action* that set out a framework for a European-Union wide policy on adaptation. The White Paper acknowledged the adaptation processes in place, but recommended that a more strategic approach be taken to integrate adaptation into existing policies (for example, agriculture, fisheries, water and biodiversity) and to ensure coherency across different industry sectors and levels of governance. The recommendations are summarized in Table 4.3 (Commission of the European Communities, 2009).

Overall, the White Paper concluded that many existing policies are well equipped to integrate adaptation measures, and set out a role for the EU to facilitate adaptation and resilience (Verschuuren, 2013). However, research concluded that this conclusion is overly optimistic, and that much greater efforts are needed to facilitate adaptation (Verschuuren, 2010). In April 2013, the European Commission adopted the EU Strategy on Adaptation to Climate Change, which sets out a framework for ensuring the preparation of the EU to current and future climate impacts. The strategy focuses on three main objectives (European Commission, 2013):

1.   Promoting action by Member States by encouraging them to adopt comprehensive adaptation strategies.
2.   Climate-proofing action at the EU level, by promoting adaptation in vulnerable sectors.
3.   More informed decision-making by addressing gaps in knowledge about adaptation and developing the European adaptation platform (Climate-ADAPT).

However, the EU Strategy on Adaptation to Climate Change does not include a proposal for an Adaptation Directive, which would mandate EU Member States to develop national adaptation policies. Only non-binding guidelines will be in place.

Canada
In November 2011, the Canadian government announced that it would invest $148.8 million over five years (2011–16) as part of a Federal Adaptation Program to support federal adaptation programs with the aim to provide credible and scientifically valid information to support adaptation planning and more informed decision-making (Natural Resources Canada, 2014). In 2011, the government released a Federal Adaptation Policy Framework, the aim of which was to enable the government to take account of climate risks when making decisions. In addition, the Climate Change Impacts and Adaptation Division (CCIAD) of Natural Resources Canada have also established an Adaptation Platform that brings together government, industry and professional groups to collaborate on adaptation priorities.

Table 4.3   Adaptation to climate change in Europe

Area	Recommendations of the 2009 White Paper
Health and social policies	– Develop guidelines and surveillance mechanisms on the health impacts of climate change by 2011. – Step up existing animal disease surveillance and control systems. – Assess the impacts of climate change and adaptation policies on employment and on the well-being of vulnerable social groups.
Agriculture and forests	– Ensure that measures for adaptation and water management are embedded in rural development national strategies and programs. – Consider how adaptation can be integrated into the three strands of rural development. – Examine the capacity of the Farm Advisory System to reinforce training, knowledge and adoption of new technologies that facilitate adaptation. – Update forestry strategy and launch debate on options for an EU approach on forest protection and forest information systems.
Biodiversity, ecosystems and water	– Explore the possibilities to improve policies, and develop measures that address biodiversity loss and climate change in an integrated manner to fully exploit co-benefits and avoid ecosystem feedbacks that accelerate global warming. – Develop guidelines and a set of tools by the end of 2009 to ensure that the River Basin Management Plans are climate-proofed. – Ensure that climate change is taken into account in the implementation of the Floods Directive. – Explore the potential for policies and measures to boost ecosystem storage capacity for water in Europe. – Draft guidelines by 2010 on dealing with the impact of climate change on the management of Natura 2000 sites (special protection areas).
Coastal and marine areas	– Ensure that adaptation in coastal and marine areas is taken into account in the framework of the Integrated Maritime Policy, in the implementation of the Marine Strategy Framework Directive and in the reform of the Common Fisheries Policy. – Develop European guidelines on adaptation in coastal and marine areas.
Production systems and physical infrastructure	– Take account of climate change impacts in the Strategic Energy Review process. – Develop methodologies for climate-proofing infrastructure projects. – Explore the possibility of making climate impact assessments a condition for public and private investment. – Assess the feasibility of incorporating climate impacts into construction standards.

Natural Resources Canada funds research on climate change impacts and provides resources to enable decision-makers to consider adaptation. These resources include adaptation-planning guidelines for local governments, communities and owners and designers of infrastructure. The platform pools knowledge, capacity and financial resources in order to determine shared priorities and address them effectively, to address issues that cross jurisdictions and affect multiple sectors of the economy to use resources efficiently, and to share data, expertise and experience thus avoid unintended negative consequences of adaptation. The objective of the Adaptation Platform is to 'create an enabling environment for adaptation, where decision-makers in regions and key industries are equipped with the tools and information they need to adapt to a changing climate' (Natural Resources Canada, 2014).

Australia
From the late 2000s onwards, the Australian government started to introduce climate adaptation strategies. However, the Abbott government, which took office in September 2013, took steps to dismantle climate change programs and introduced legislation to repeal the Emissions Trading Scheme. The future of climate action in Australia is therefore uncertain.

In 2007, the Council of Australian Governments (COAG) agreed to the National Climate Change Adaptation Framework. The framework 'outlines the future agenda of collaboration between governments to address key demands from business and the community for targeted information on climate change impacts, and to fill critical knowledge gaps which currently inhibit adaptation' (Department of Climate Change and Energy Efficiency, 2007). The framework acknowledged the *ad hoc*, fragmented and limited nature of research in adaptation and impact of climate change, and suggested the following areas for reform (Department of Climate Change and Energy Efficiency, 2007):

- Establishment of an Australian Centre for Climate Change Adaptation.
- Establishment of a national program to develop and communicate information relating to regional climate change and vulnerability.
- Development of integrated regional vulnerability assessments.
- Improvement in communication, information and tools to allow more informed decision-making.
- Development of international connections and partnerships to allow Australia to contribute knowledge and learn from the experience of other nations.

The federal government responded proactively to some of these recommendations, establishing a CSIRO National Research Flagship in Climate Adaptation in 2008 and the National Climate Change Adaptation Research Facility (NCCARF) in 2008. The National Climate Change Adaptation Framework also put forward recommendations in order to reduce vulnerability of particular sectors and regions. The recommendations are summarized in Table 4.4 (Department of Climate Change and Energy Efficiency, 2007).

Table 4.4   Australia's National Climate Change Adaptation Framework

Area	Recommendation to reduce vulnerability of area
Water resources	– Research to address knowledge gaps regarding climate change and water resources. – Work with the water industry to ensure that climate change impacts and risks are incorporated into water resource and infrastructure planning and management strategies.
Coastal regions	– Undertake a comprehensive assessment of Australia's coastal vulnerability. – Apply appropriate planning policies for vulnerable coastal areas.
Biodiversity	– Review the National Biodiversity and Climate Change Action Plan 2004–07. – Implement a national program to understand the impacts of climate change on biodiversity. – Provide practical guidance on how to integrate existing and new information about climate change into management of disturbance regimes in areas managed for biodiversity conservation. – Assess the vulnerability of Australia's World Heritage and other internationally important properties to the impact of climate change. – Implement steps in the climate change action plan for the Great Barrier Reef.
Agriculture	– Implement adaptation components of the National Agriculture and Climate Change Action Plan.
Fisheries	– Develop a climate change and fisheries action plan. – Address knowledge gaps about the impact of climate change on wild catch fisheries and aquaculture.
Forestry	– Develop a climate change and forestry action plan. – Address knowledge gaps about the impact of climate change on forestry and the vulnerability of forest systems.
Human health	– Australian Health Ministers's Conference to implement a National Action Plan on Climate Change and Health. – Develop and implement heat wave warning and response systems. – National Health and Medical Research Council to increase its focus on research on climate change and health. – The Sport and Recreation Ministers' Council to develop and implement an action plan to assess and develop strategies to address the impacts of climate change on sporting and recreational activities.
Tourism	– Tourism Ministers' Council to develop an action plan in partnership with industry stakeholders to assess the impacts of climate change on tourism and tourism values.

Area	Recommendation to reduce vulnerability of area
Settlements, infrastructure and planning	– Identify and address knowledge gaps and promote synthesis of existing information. – All jurisdictions to promote decisions that increase resilience to climate change and discourage decisions that increase vulnerability and consider changes where appropriate. – Analyze and revise planning systems to increase resilience to climate change. – Identify and address impacts of climate change on infrastructure. – Develop a partnership with the insurance and finance industries. – Establish a program to assist local government in adapting to climate change.
Natural disaster management	– Undertake research to improve knowledge on the nature, and expected extent of changes to existing risk profiles (for example, bushfires or flooding) as a result of climate change. – Incorporate climate change impacts into planning or natural disaster response management.

The Australian government implemented some of these recommendations. For instance, the Department of Agriculture produced a National Climate Change and Commercial Forestry Action Plan in 2009, and a National Climate Change Action Plan for Fisheries and Aquaculture in 2010. NCCARF developed a National Climate Change Adaptation Research Plan in a number of areas, including biodiversity, health, settlements and infrastructure, water, and disaster management and emergency services. In 2009, the House of Representative Standing Committee on Climate Change, Water, Environment and the Arts published a report on the inquiry into climate change and environmental impacts on coastal communities. The report included 47 recommendations aimed at improving national leadership to promote sustainable use of Australia's coastal zone, titled *Managing Our Coastal Zone in a Changing Climate: The Time to Act is Now* (House Standing Committee on Climate Change Water Environment and the Arts, 2009). In the same year, the Department of Climate Change published a national assessment of climate change risks to Australia's coast, titled *Climate Change Risks to Australia's Coasts* (Department of Climate Change, 2009). The objectives of the report were to provide an initial assessment of future implications of climate change for significant aspects of Australia's coast, as well as to identify areas at high risk to climate change impacts, key barriers to minimizing impacts of climate change, and national priorities for adaptation to reduce climate change risk in the coastal zone (Department of Climate Change, 2009). In response to these reports, the Commonwealth government sponsored a national coastal climate change forum in late 2010 (Department of Climate Change and Energy Efficiency, 2010).

In 2012, the Commonwealth government released a discussion paper that outlined principles for allocating management of climate change risks, roles and responsibilities for adapting to climate change within the three levels of government: Commonwealth, state (or territory) and local (Department of the Environment, 2014). The report also set out a number of guiding principles for the management of climate change risks, including recommendations on how governments should respond to market and regulatory failures that prevent effective and efficient climate change risk management. In summary, the report endorsed a system based on private responsibilities and local initiatives, with the Commonwealth government having a minor role to play (Department of the Environment, 2014).

Most recently in 2012, the Productivity Commission initiated an inquiry into barriers to effective climate change adaptation, with a final report being published in March 2013 (Productivity Commission, 2012). While the report supported the view of the government in relation to private responsibilities for adaptation, it adopted a broader view in relation to the responsibility of government bodies, stating that they have a role to play where climate change poses risks to the activities of government bodies, where goods and services necessary to bring about adaptation are underprovided by the private market and where regulatory and policy frameworks are necessary to manage adaptation decisions. Whether and how the proposed policy reforms (including tax incentives, changes to regulatory arrangements as well as the roles, responsibility and liability of local government) will be implemented will have to be seen in the future.

Adaptation and resilience on local levels
Irrespective of international and national progress on achieving a global climate change outcome, the importance of both adaptation and resilience to climate change is increasingly recognized, especially in sectors and locations that have a high vulnerability to climate change and/or a long history of exposure to changing environmental conditions. This recognition is often driven by the experience of local vulnerabilities (for example, natural disasters such as flooding, or hurricane/cyclone impacts), or by concerns about future changes in climate and implications for local regulations and laws, as in the case of sea level rise. Local governments and councils are also experiencing the need to consider adaptation and resilience. Byron Shire Council on the East Coast of Australia was exposed to litigation following their 'planned

retreat' policy allowing the shoreline to advance inward without protection of buildings and other infrastructure. New York City organized the New York City Panel on Climate Change (NYPCC) in 2008 as part of its PlaNYC, the city's long-term sustainability plan,[2] leading to one of the more advanced adaptation plans in existence. The New York Academy of Science, which published the plan, stated that:

> Climate change has the potential to affect everyday life in New York City. Environmental conditions as we experience them today will shift, exposing the city and its residents to new hazards and heightened risks; we will be challenged by increasing temperatures, changes in precipitation patterns, rising sea levels, and more intense and frequent extreme events. While mitigation actions that reduce greenhouse gas emissions will help to decrease the magnitude and impact of future changes, they will not prevent climate change from occurring altogether.

After experiencing the impacts of hurricanes Irene and Sandy, New York City developed a resilience plan titled *A Stronger, More Resilient New York*, with recommendations both for rebuilding the communities impacted by Sandy and increasing the resilience of infrastructure and buildings citywide. While the resilience plan has not remained without criticism (in particular as it made an argument for the increased resilience of fossil fuel infrastructure), it is one example for current action that is already being undertaken. Similar plans and programs are emerging in orther regions and localities. For example, the local government in New South Wales (Australia) has implemented a $1 million 'Building Resilience to Climate Change' program for projects that build resilience to extreme heat, water supply shortages or adapt priority infrastructure to climate change impacts.

While some initiatives are taking place, stringent efforts to adapt and build resilience across organizations, industry and society are not yet evident. In the next section, we look as those responses that can be undertaken at organizational levels in order to design and/or implement measures to facilitate adaptation and resilience to climate change.

Concluding Comments

This chapter has outlined how international climate policy has been translated into national and sector-specific measures. The first part of the chapter has provided an overview of different approaches taken towards

---

[2] Accessed 22 September 2014, available at http://www.nyas.org/publications/annals/Detail.aspx?cid=ab9d0f9f-1cb1-4f21-b0c8-7607daa5dfcc.

the mitigation of carbon emissions. The second part of the chapter has looked into the emerging policy framework around adaptation, and has examined approaches taken by LDCs and developed countries on adaptation to climate change. The chapter has thereby placed a focus on policy frameworks that have been used to guide action on climate change (both in terms of mitigation and adaptation) and to facilitate action on climate change more widely. Despite the emergence of various policy approaches aimed at mitigation, progress has been slow, and it can be concluded that no stringent inter-governmental measures are in place to halt carbon emissions beyond regional initiatives, such as those in the European Union.

With carbon emissions on the rise, the topic of adaptation is becoming more important. However, regional adaptation efforts have been slow and limited, with UNFCCC efforts primarily driving LDCs to draft their NAPAs. International approaches have been of varying levels of development, and have different implications for private-sector development. While the Cancun Adaptation Framework and the establishment of the Adaptation Committee in 2010 have placed the topic of adaptation higher on the international agenda, most current approaches only provide high-level guidance to governments, communities or organizations on adaptation options and managing climate-related risks, without putting any stringent or legislative adaptation requirements in place. Consequently, communities and private and public sector organizations will be required to develop responses to impacts in the absence of guiding policies or frameworks.

PART II

Organizational responses

5. Vulnerabilities and impacts as drivers for change

I suppose the world is sitting back and waiting for someone to come up with a better strategy

Sustainability manager, rail company

Awareness of the need for adaptation has slowly been growing in international climate policy developments (as outlined in the previous chapters), but also in risk management, disaster planning and private sector management. Given the rising levels of greenhouse gas (GHG) emissions as well as the considerable inertia in achieving stringent mitigation responses, adaptation will be required to address the adverse impacts from climate change. Research increasingly suggests that many manifestations of climate change will be localized, and will require coordinated, planned and context-specific action. Some levels of adaptation will be necessary to address those impacts from climate change that would occur even if stringent cuts of GHG emissions were implemented on a global scale immediately. As outlined in Chapter 2, the latest assessment report of the Intergovernmental Panel on Climate Change (IPCC) concluded that: '[m]ost aspects of climate change will persist for many centuries even if emissions of CO_2 are stopped. This represents a substantial multi-century climate change commitment created by past, present and future emissions of CO_2' (IPCC, 2013). The previous chapter has demonstrated that there is now growing attention to adaptation within the United Nations Framework Convention on Climate Change (UNFCCC), but also by national policy-makers and civil society. Adaptation has emerged as an important pillar of international discussions on a future climate regime under the UNFCCC, alongside mitigation, technology and finance (UNFCCC, 2010).

In addition to the growing awareness of the need for adaptation, there is now also a growing realization that gradual adaptation to climate change alone might not be sufficient to cope with some of the more extreme outcomes of climate change, such as changes in the frequency and/or severity of weather extremes. Consequently, arguments have been put forward that organizations, industries and society as a whole need to also strengthen their resilience, that is, the capacity to absorb and withstand adverse impacts, and to recover from more drastic environmental changes beyond the circumstances that an organization is adapted to and can cope with (Linnenluecke and Griffiths, 2010). A key concern is that organizational adaptation may not provide a sufficiently strong response to cope with more disruptive and abrupt changes as a result of climate change. Overall, adaptation may be more appropriate as a response mechanism for adjusting to longer-term and more gradual climate change trends (for example, gradual temperature rise) (Linnenluecke and Griffiths, 2010).

Despite the growing realization that climate change impacts will pose significant challenges, the majority of businesses and other organizations, such as those from public sectors, have repeatedly been criticized for their lack of responses and adaptation to the physical impacts of climate change. One of the clearest conclusions about organizational adaptation and resilience to climate change and weather extremes is that little progress has been made towards developing and implementing decisive adaptation plans – irrespective of whether public or private actors are involved (Linnenluecke, 2013a). There are only a few actual examples (besides those from exposed organizations and sectors, and those from a few leading companies) that demonstrate stringent actions, in particular across government, communities, industries, or groups of businesses. Corporate action on any climate-related issues is often initiated retrospectively or to respond to short-term fluctuations only. For instance, in fisheries and aquaculture, coping with climate variability is 'business-as-usual' for many companies, but rarely includes long-term anticipatory adaptation as a preparedness strategy for climate change and a greater level of climate variability (Hobday et al., 2012).

To date, discussions about organizational responses to climate change have largely focused on an 'environmental impact' debate, rather than on an environmental response debate (that is, the impact of human activities *on* the environment, rather than vice versa), which has also been driven by underlying policy developments (Winn and Kirchgeorg, 2005; Winn et al., 2011). To some extent, the lack of action can also be attributed to mismatches between the occurrence of climate change and organizational decision-making. Organizational decision-makers are typically focused on short-term planning horizons and outcomes, rather than on longer-term adaptation planning needs. Other factors hindering progress on adaptation are uncertainties surrounding future risks, the availability of data and models on a local level, the availability of capabilities and capacities, unclear problem definitions and risk-assessment methodologies, unclear legislative frameworks and planning mechanisms, and an overall lack of political leadership and community support (IPCC, 2014c). Furthermore, climate adaptation and mitigation are often pursued as separate activities, with little integration among the two.

This chapter looks at challenges directly associated with impacts and vulnerabilities (that is, the predisposition to be adversely affected (IPCC, 2012)) as they affect organizations and create challenges to adapt and the need to create resilience. We review traditional and emerging drivers for organizations to respond to a changing environment, and put forward a framework for evaluating both impacts and vulnerabilities. Taken together with the significant uncertainties surrounding the impacts of climate change, it is evident that much remains to be learned about 'what actually works' when it comes to adaptation (McGray, 2013), and that successful adaptation will not be easy to achieve (Linnenluecke, 2013a). A great challenge related to adaptation is that outcomes are difficult to assess due to cost-benefit asymmetries (that is, near-term investments are not leading to near-term outcomes) and due to the consequences of the types of impacts that an organization cannot adapt to easily (which will thus need resilience capacities). We will further discuss these issues in the next chapter.

Traditional Drivers for Organizational Change

Traditionally, the question of how organizations should respond actual or predicted environmental changes has focused on changes in the business context or the organization's external environment, with relatively few insights into organizational responses to natural environmental change (Linnenluecke and Griffiths, 2010). In many industries, the drivers of competitive success have been process improvements and innovation – both technological and organizational – in response to changing institutional or regulatory conditions, a changing competitive landscape, or other types of political, economic, social, demographic and technological changes (Linnenluecke et al., 2012). From the latter part of the twentieth century, many organizations faced significant pressure from globalization as a new challenge in their environment which resulted in declining trade barriers and significant increases in global competition. Managers have been attempting to adapt their organizations to survive and grow in these changed circumstances. Organizations have also encountered significant levels of technological and social change. The advances in information technology have been driving technological innovations, which has radically changed communication, shortened product life cycles and speeded-up communication channels from suppliers, to manufacturers, service providers and customers.

In addition, social changes have occurred, leading to greater social differentiation, heightened customer expectations and a shift of demand from mass production to 'mass customized' or fully customized products (Dunphy and Griffiths, 1998; Dunphy et al., 2003). Organizations, therefore, have faced increasing pressures to adopt flexible manufacturing or service delivery technologies due to the need to increase flexibility, productivity, and – at the same time – the quality of manufacturing or service delivery processes. In many instances, organizational changes to meet external challenges have been transformative – organizations were downsized or redesigned around strategic units rather than functional departments, and the main focus rested on creating and maintaining high levels of performance. These organizational shifts also created a new demand for creative and knowledge-based people. Only few organizations were exempt from feeling pressures for changes – such as those in sheltered industrial niches or government sectors where changes were more incremental with moderate attempts at being more responsive, adaptable and flexible towards external environmental change.

Emerging Drivers for Organizational Change

While organizations have a long history of adaptation to changing industry structures and institutional conditions, the natural environment and physical impacts of the natural environment on organizations have usually not been included in dominant frameworks, and little consideration has been given to temporal and spatial variances in natural systems (Linnenluecke et al., 2012). However, and as outlined in Chapters 1 and 2, global environmental change and climate change bring about a host of new challenges for organizations, requiring them to take into consideration the dependency on climate-sensitive or non-renewable natural resources, emerging climate legal risks, vulnerabilities to the physical impacts of climate change, changes in investment strategies, insurance affordability and availability, stakeholder pressures, as well as more systemic risks within the economy and supply chains. There are significant technological advances which could ultimately lead to a post-industrial age and significant shifts in how the economic system uses national and social capital to support financial and productive capital (Senge and Carstedt, 2001).

Some exposed companies such as those in the reinsurance industry (for example, Munich Re, Swiss Re) have drawn the links between climate change and impacts on their business models. These companies have started to undertake research into the risks associated with a changing climate and impacts on their organizations. Topics of interest include the viability of current insurance models, changes to patterns of extreme weather events and resulting losses for the reinsurance industry, but also insurance models for renewable energy projects, as well as the need to consider legal aspects to do with climate impacts on the reinsurance industry. Given the severity of the climate projections, other major companies are slowly becoming concerned that they are at risk from climate change and impacts from extreme weather events. The latest Carbon Disclosure Project (CDP) Global 500 report suggested that 43 per cent of companies reported 'changes in precipitation extremes and droughts' as climate risks, up from just a handful of companies a few years ago.

CASE STUDY: CLIMATE CHANGE AS AN EMERGING DRIVER FOR CHANGE IN THE INVESTOR COMMUNITY

A study by Carbon Tracker and the Grantham Research Institute on Climate Change and the Environment at the London School of Economics and Political Science (LSE) in the UK has highlighted risks for regulators, governments and investors stemming from climate change. Their calculations suggest that about 60–80 per cent of coal, oil and gas reserves of publicly listed companies are 'unburnable' if governments are to meet global commitments to keep the temperature rise below the so-called 2°C target. The possibility that coal, oil and gas reserves may become 'stranded' and associated investments 'wasted capital' has serious implications for institutional investors and the valuations of companies with fossil fuel reserves.

When looking at the adaptation challenge, two separate issues come together (see Figure 5.1) – one issue is associated with the level of physical impacts of climate change (which are dependent on underlying drivers such as world development and future levels of GHG emissions), the other is associated with the level of vulnerability of a certain region, sector or industry, which can be understood as the 'function of the character, magnitude, and rate of climate variation to which a system is exposed, its sensitivity, and its adaptive capacity' (IPCC, 2001). In other words, vulnerability can be understood as the predisposition to be adversely affected (IPCC, 2012). Different regions, sectors or industries have different levels of adaptive capacity, defined as: '[t]he combination of the strengths, attributes, and resources available to an individual, community, society, or organization that can be used to prepare for and

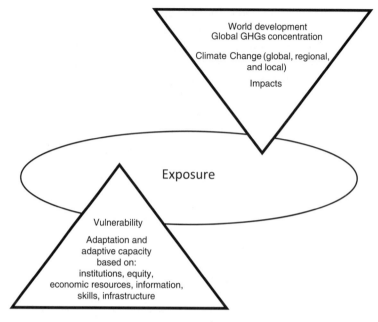

Source: Adapted from Dessai and Hulme (2004).

Figure 5.1   The climate change challenge

undertake actions to reduce adverse impacts, moderate harm, or exploit beneficial opportunities (IPCC, 2012: 556)'. The connecting factor between the physical impacts of climate change and vulnerability is exposure, defined as: 'the presence of people; livelihoods; environmental services and resources; infrastructure; or economic, social, or cultural assets in places that could be adversely affected' (IPCC, 2012: 559).

Understanding Vulnerability: the Coping Range of Organizations

Adaptation is needed to respond to the adverse effects of climate change that are either already observable or projected to occur in the future. Adaptation enables organizations to minimize vulnerabilities to the impacts of climate change by undertaking changes to their so-called 'coping range' (Smit et al., 2000; Yohe and Tol, 2002). The idea of a coping range is rooted in Hewitt and Burton's (1971) regional ecology of hazards and the concept of vulnerability, and was subsequently adopted by studies looking into climate vulnerability (Adger et al., 2003; Smit and Wandel, 2006; Smith et al., 2000; Yohe and Tol, 2002), including the

IPCC Third Assessment Report. In these studies, the coping range is defined as the capacity of social or economic actors to accommodate variations in climatic conditions. Applying the concepts to organizations, a coping range can be understood as variations in one or more climate-related variables such as temperature or rainfall that an organization can tolerate without experiencing any adverse consequences (Linnenluecke and Griffiths, 2011, 2012).

Most organizations can tolerate at least some variability in climatic conditions such as seasonal or inter-annual changes (Linnenluecke and Griffiths, 2012). Such variability falls within the core of the coping range and is not leading to any adverse consequences. Towards one or both thresholds of the coping range, the climatic conditions are are not ideal for the organization, but still acceptable. The thresholds of a coping range indicate boundaries beyond which adverse consequences become visible, such as losses in production yields if growing conditions are negatively impacted by a heat wave or above average rainfall (Carter et al., 2007; Linnenluecke and Griffiths, 2010, 2012). Especially organizations in ecosystem-dependent industries (for example, agriculture, viticulture, fisheries, forestry, and pastoral industries) are highly reliant on certain temperature and rainfall patterns. These organizations typically experience beneficial outcomes when operating under conditions that match these patterns, but are challenged by conditions that deviate, such as lasting changes in average temperature (Linnenluecke and Griffiths, 2012).

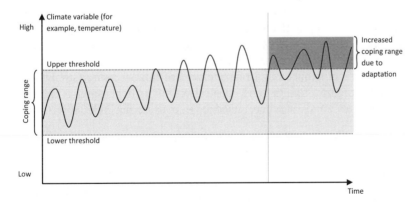

Sources: Adapted from Carter et al. (2007) and Lemmen and Warren (2004).

Figure 5.2   The coping range of an organization

Vulnerabilities for organizations arise as climate change increases the probability that conditions outside the boundaries of the coping range occur. For instance, an individual climate variable such as temperature or rainfall might deviate from 'usual' conditions as the climate changes. If such changes are long-lasting and persistent, and if the coping range is narrow, an organization will experience adverse impacts (Linnenluecke and Griffiths, 2012). For instance, a study of Australia's 61 recognized wine regions found a median growing season temperature of over 21°C (an indicator of the limit of quality wine grape production conditions) for three regions for the period 1971–2000, for eight regions for a 2030 climate change scenario, for 12 regions for a 2050 scenario, and for 21 regions for a 2070 scenario (Hall and Jones, 2009). A coping range can also be exceeded by increases in the frequency and/or intensity of weather extremes under a changed climate (such as droughts or floods), which are more difficult to adapt to as they occur infrequently.[1] Yohe and Tol (2002) used the coping range to investigate hypothetical upper and lower critical thresholds for the River Nile under future climate change. The researchers based their study on current and historical stream flow data. The historical frequency of exceeding the coping range served as a baseline from which to measure changing risks using a range of climate scenarios. Based on their findings, Yohe and Tol defined upper thresholds (serious flooding) and lower threshold (minimum flow required to supply water demand) which allows decision-makers to understand future flood risks (Carter et al., 2007).

Broadening the coping range through adapting

Organizations can undertake adaptation measures to widen their coping range, for example, by developing drought resistant crops or flood barriers. A wide variety of different adaptation options can be identified, falling into three general categories: *structural/physical options* (improved maintenance and management of infrastructure, strengthening of infrastructure foundations, protection of critical assets, enhancing redundancies, and – where needed – considering relocation), *social*

---

[1] It should be noted here that many extremes occur naturally even under an unchanged climate and that not all extremes are undesirable. In some cases, extremes such as fires, droughts or flood are important to the health of ecosystems. However, any increases in the frequency and intensity of weather extremes can easily exceed existing coping capacities and lead to significant damages (Linnenluecke, M.K. and Griffiths, A. (2012). 'Assessing organizational resilience to climate and weather extremes: complexities and methodological pathways', *Climatic Change*, **113**(3–4): 933–47.

options (raising awareness and education, improving information), and *institutional options* (incentives through taxes or subsidies, insurance, zoning laws and building standards, national and regional adaptation plans, improved disaster planning and forecasting) (Noble et al., 2014). In addition, other criteria can be applied to classify adaptation. These include the type of *adaptation stimuli* (which can be climate change or other types of external risks and opportunities), *the type of response* (reactive or anticipatory), as well as the *planning horizon* (spontaneous/ autonomous or planned) (Smit et al., 1999). Autonomous adaptations are those taking place without a directed plan or intervention. Such adaptations may not be optimal if there are constraining factors such as a lack of integration of actions across different actors, uncertainties about climate impacts, imperfect information, and high costs of implementing adaptation decisions. Adaptation can refer to both the process of adapting and to the resulting outcome (European Commission, 2009; Smit et al., 1999).

Felgenhauer and Webster (2013) proposed to disaggregate adaptation into flow adaptation responses (investments that are relatively low cost, quick to implement and to return benefits, and limited in the level of damages that can be avoided), committed adaptation stock (investments that are long-lived and depreciable, which typically includes large-scale and high-cost infrastructure), and option stock (an initial investment in preparatory adaptation with a lower adaptive capacity that enables easy expansion in the future, if needed to address rising climate damages). This approach can be used in an organizational setting to assess which of the three categories the adaptation method fits into, also considering options for integrating or supplementing different adaptation measures with mitigation (Felgenhauer and Webster, 2013).

In many cases, successful adaptation will depend on responses to local manifestations of global changes, supported by the integration of measures across jurisdictional, administrative and geographical scale, such that adaptation does not occur in isolation. Climate adaptation in industry and business (especially technological adaptation options) are often motivated by market signals, which may at times not be well-matched with adaptation needs and residual uncertainties (Revi et al., 2014). For example, adaptations that primarily create public goods (coastline protection) might not receive private-sector funding. In cases where adaptation primarily benefits broader society over private actors, it might thus require public action, for example, by local governments or NGOs (Chambwera et al., 2014). Adaptation is also not isolated from other decisions, but occurs in the context of ongoing political, economic, social and technological changes. Nonetheless, irrespective of the motivation, adaptation can generate short-term or long-term benefits if they align well with environmental changes (Adger et al., 2005).

In addition, it should be noted that organizations depend on larger systems within which they are embedded. For instance, individual businesses and industries are dependent on products, services, infrastructure and building stock managed by other organizations. They are also dependent on public sector infrastructure, including transport, drainage, electricity transmission systems, public water supply and treatment. Adaptation therefore depends on adjustments in these systems, as well as on changes in regulatory and governance frameworks such as zoning and planning regulations (Noble et al., 2014). The systems perspective is also important for understanding resilience – organizations and industries embedded in regions or localities which have strong financial, governance or emergency response systems are typically more resilient than organizations and industries without such support (see Chapter 7). National adaptation plans are likely to include preferences towards certain measures from structural, institutional and social options, which may enable or constrain the adaptation in certain sectors. The case of Eskom (see Chapter 3) shows how companies can take more active roles in adaptation policy-making. Eskom provided written inputs to the South African government's National Climate Change Response White Paper, and engaged in international policy dialogues on climate change adaptation (United Nations Global Compact and United Nations Environment Programme, 2012).

Given the embeddedness of organizations within larger systems, the action of individual organizations will likely not be enough to create resilient industries and communities. Adaptation will require coordinated action between organizations, and between organizations, communities and policy-makers. Researchers have already started to study different governance structures such as collaborative networks among private and public organizations, and how these structures are likely to influence adaptation on industry-levels in the face of climate change (Wyss et al., 2014). Importantly, the adaptive capacity of any business or industry sector is also largely influenced by its socio-economic environment and factors at different scales (see Chapter 1). For instance, the adaptive capacity of small, family-owned enterprises in less-developed countries (LDCs) is constrained by poor institutional support, lower levels of education, water stresses or biodiversity loss, in addition to other factors such as poverty, unemployment, political and/or social instability, and land degradation (IPCC, 2014b). In this and the next chapter (Chapter 6), we further explore issues around adaptation, as well as costs-benefit considerations.

Losses in resilience if the coping range is exceeded

Organizations can implement adaptation measures to broaden their coping range, such as the development of drought resistant crops or the implementation of flood barriers. Adaptation timeframes typically range from several months to several years. Any changes in climate or extreme weather events that occur with greater magnitude, frequency or abruptness than expected can, therefore, easily exceed the coping range and thresholds for adaptation. In these situations, organizations might not be able to implement adaptation measures fast enough, or existing adaptation measures might be insufficient to alleviate impacts. Similar issues can arise when impacts are more persistent than expected, for instance, a long drought might prove challenging beyond a certain point (Linnenluecke and Griffiths, 2012). In these cases, an organization can experience a loss of resilience, depending on its capabilities to respond to more unexpected, severe or extreme changes.

The figure below illustrates two hypothetical scenarios (gradual changes in average temperatures due to global warming as well as the impact of a hurricane) that show how climate change and weather extremes can impact organizational performance once the coping range is exceeded. Corporate performance can be measured in terms of financial performance or output, such as agricultural production. In extreme cases, exceeding the coping range can lead to organizational decline and eventual failure. Indicated in this figure is therefore a threshold beyond which an organization cannot recover from adversity, represented by the dashed horizontal line (Linnenluecke and Griffiths, 2012). Statistics already show that the impacts of extreme events are significant on small businesses – according to US data, 25 per cent of businesses do not reopen after a natural disaster. Even more staggering is that 75 per cent of businesses fail within three years of a natural disaster if they did not have a business continuity plan in place (Impact on US Small Business, 2007: 4). These statistics are likely to grow if impacts of the extreme event are becoming more noticeable over time.

Figure 5.3 shows also the relationship between adaptation and resilience. A more adapted organization is less susceptible to adverse impacts in the first place. However, once the coping range is exceeded and impacts become more noticeable, those organizations that possess resilience are either less impacted by these impacts initially (assuming that all other factors are equal), and/or are able to recover faster than other organizations that are less resilient. We further discuss resilience mechanisms in Chapter 7.

Of particular concern is that umitigated climate change might ultimately lead to abrupt climate change over entire regions (Alley et al.,

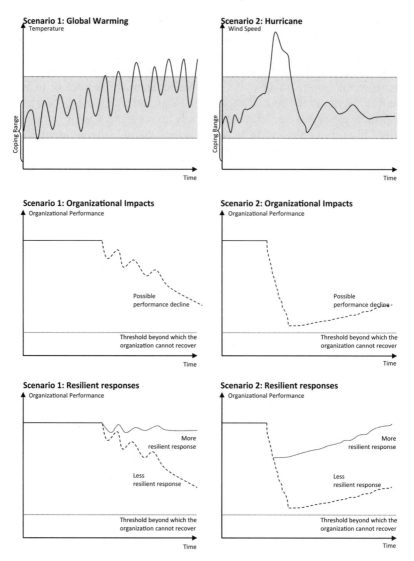

Source:　Adapted from Linnenluecke and Griffiths (2012).

Figure 5.3　The coping range under different climate change scenarios

2003), which could lead to catastrophic impacts across scales if they were to occur (Linnenluecke and Griffiths, 2012; Wilbanks et al., 2007b). Abrupt climate change includes:

both the abrupt changes in the physical climate system ('abrupt climate change') and abrupt impacts in the physical, biological, or human systems triggered by a gradually changing climate (hereafter called 'abrupt climate impacts'). These abrupt changes can affect natural or human systems, or both. The primary timescale of concern is years to decades. A key characteristic of these changes is that they can unfold faster than expected, planned for, or budgeted for, forcing a reactive, rather than proactive mode of behavior. These changes can propagate systemically, rapidly affecting multiple interconnected areas of concern (National Research Council, 2013: 27).

These include possible abrupt changes such as sea level rise from a destabilization of the West Antarctic Ice Sheet, changes to patterns of climate variability, increases in the intensity, frequency and duration of heat waves and other extremes or rapid changes in ecosystems (National Research Council, 2013).

Such large-scale impacts may not be as unlikely as previously thought. A recent study by NASA and researchers at the University of California, Irvine, found that a rapidly melting section of the West Antarctic Ice Sheet appears to be in an irreversible state of decline. The study presents multiple lines of evidence based on 40 years of observations. Findings from the study indicate the glaciers in the Amundsen Sea sector of West Antarctica 'have passed the point of no return'. These glaciers already contribute significantly to sea level rise as they release annually almost as much melting ice into the ocean as the entire Greenland Ice Sheet. It is estimated that these ice sheets contain enough ice to raise global sea level by 4ft (1.2m) if they were melting. A conservative estimate is that it might still take several centuries for all of the ice to flow into the sea (thus leaving time for adaptation), but melting is already occurring faster than expected. As the retreat of the ice sheet is happening simultaneously, the authors argue that this is triggered by a common cause, such as an increase in the amount of ocean heat beneath the floating sections of the glaciers.

Climate change and supply chain risks
In addition to vulnerabilities within their own value chain, organizations face considerable risks that climate change poses to their supply chains. Supply chains are a useful construct for examining industrial production in its entirety as it represents how individual organizational activities are assembled to provide an effective product or service delivery beyond the individual organizational activity (Fleming et al., 2014). Supply chain risks range from the impacts of extreme weather events on transportation infrastructure, disruptions in an organization's supplier base due to reduced or interrupted production capacity, changes in the demand for

goods, and even the inability to do business in certain regions. The impacts of Russia's export ban that was imposed in 2010 highlights such risks. At the time, Russia experienced a severe heat wave and large-scale wildfires that destroyed crops, particularly wheat. As news of this disaster became known, and as the impacts of the resulting drop in Russia's grain crop were felt among international supply chains, international grain prices increased dramatically. In an attempt to stabilize domestic grain crop prices, Russia's prime minister Vladimir Putin announced a ban on grain exports which pushed wheat prices to a 23-month high on commodities markets (Parfitt, 2010). While this example is from the agricultural sector, similar examples can be drawn from other sectors that are dependent on physical resource inputs, such as the mining, water or the energy sectors.

Analyzing and understanding the risks that climate change introduces across supply chains is not an easy task, as information is needed about the coping range of actors across the supply chain as well as so-called climate 'hot spots'. These are countries or regions where climate change impacts are likely to be particularly pronounced (Giorgi, 2005). However, these 'hot spots' have typically not been fully integrated into supply chain management and risk analysis systems. The carbon footprint of supply chains will become increasingly relevant in view of changing mitigation legislation. However, conventional approaches to analyzing supply chain risks often fail to address the environmental (and also social risks) within global supply chains.

In addition to the impacts of climate change, unsustainable supply chain practices (such as the overuse and over-exploitation of natural resources in vulnerable areas) can destroy biodiversity, deplete freshwater resources, lead to changes in land-use, or contribute to rising chemical pollution (Rockström et al., 2009b). Unsustainable supply chain practices can also have negative impacts on local or indigenous communities through improper consultation or low health and safety standards. Ultimately, such unsustainable practices can erode an organization's production base and also threaten an organization's reputation. The Rana Plaza building collapse in Bangladesh, an eight-story commercial warehouse used for garment manufacturing, has led to widespread discussions about working conditions and safety standards across global supply chains. Several international clothing brands were among the companies using the warehouse for the manufacturing of clothes. This examples shows that supply chain risks are related to a wide-ranging number of factors which often exceed climate change alone and relate to broader sustainability issues.

Understanding Impacts: What Future are Businesses Preparing for?

The assessment of vulnerabilities is only one part of understanding the adaptation challenge. The other issue is associated with understanding the physical impacts of climate change and the projected levels of future impacts. We are now entering an era where we have more knowledge on climate change and impacts that we ever had before. Over the last few decades, the ability to simulate the interactions of the atmosphere, oceans, land surface, and ice and to project temperature changes resulting from increases in atmospheric concentrations of GHGs has greatly improved. At the same time, knowledge about the possible range of climate change outcomes has also increased. Recent scientific models and assessments are much more advanced, showing that the Earth's climate is not behaving in a linear and predictable fashion. Standard assessments prior to 1995 modeled climatic responses to a one-time doubling of CO_2 and analyzed the effects once the system reached equilibrium. The assumption was that the probability of climate extremes such as droughts, floods, and hurricanes would either *remain unchanged* or would *change gradually* in accordance with global warming. In other words, these early models assumed that more global warming would lead to more hot days, and thus a greater probability of heat waves or droughts (Schneider, 2004). Scientists now focus on understanding the whole range of potential climate change outcomes in addition to gradual changes, including changes in weather extremes and potentially even abrupt climate change. Some changes in weather extremes have already been observed.

As both climate science and climate change are progressing, decision-makers are also faced with the issue that there is overall less certainty in regards to the consequences (Trenberth, 2010). The ability to project how climate change may affect industry, businesses and communities is limited by uncertainties about climate change itself at a fine-grained sectoral and geographical scale (Linnenluecke and Griffiths, 2012; Wilbanks et al., 2007b). The ability to project how climate change may affect industries, businesses and communities is also limited by uncertainties about vulnerabilities and the need to adapt over the next century regardless of climate change (Wilbanks et al., 2007b). This situation creates great difficulties for decision-makers. In order to implement effective adaptation policies, plans and progress, climatic and non-climatic information and data are required, also across sectors, and at local, regional, national and global levels. Climatic data encompasses systematic observations of changes in climate and weather patterns, while non-climatic data encompasses information on vulnerabilities and adaptation options. Importantly, the information needed for decision-making purposes extends to adaptation to future climatic conditions, and not just to past and present conditions (UNFCCC, 2010). Even though there are significant uncertainties around the precise consequences of climate change, there are nonetheless various valuable sources on projected climate impacts. Table 5.1 highlights some of the information available for decision-makers.

The sources listed in Table 5.1 provide different outputs relevant for different audiences. Some of the reports and projections (Gledhill et al., 2013) are directly targeted at a business audience, and can thus be easier to use for decision-making in a business context. The case study below shows how one company in the US has used the Climate Ready/UK Climate Projections as inputs into its decision-making. Some of the more specific modeling approaches are more relevant for climate-change impact modelers and policy analysts. The downscaling techniques (MAGICC/SCENGEN, SDSM) described here can be applied to generate small-scale climate data that is often used in impact models, and for the development of future climate scenarios at both local and national scales. Nonetheless, results of such models are also important to show future developments in climate to derive information about possible impacts for businesses. We have omitted models used in climate research such as the Atmosphere-Ocean General Circulation Models (AOGCMs) and Earth System Models (ESMs) as those techniques focus on global climate modeling and also require considerable scientific expertise and experience in climate modeling and climate simulations. We are also not discussing more complex Regional Climate Models (RCMs) as the book is not targeted at a specialist modeling audience. The interested reader is referred to the latest IPCC report that includes a detailed discussion, the CMIP5 simulations (Coupled Model Intercomparison Project Phase 5), and the UNFCCC compendium on methods and tools (UNFCCC Secretariat, 2005).

Assessing Organizational Adaptation Needs and Priorities

Using the impact information in the previous sections as guidance, the question arising is how decision-makers can best assess and evaluate the potential impacts of climate change on their organizations. A large number of traditional tools for strategic analysis are available for decision-makers, including the basic Political, Economic, Social and Technological (PEST) analysis and Porter's Five Forces for the analysis of external and industry trends, as well as tools such as value-chain

Table 5.1 Information for decision-makers on climate change impacts and related risks

Source	Description	Climate variables covered	Other information	Scale
Business-not-as-Usual: Tackling the Impact of Climate Change on Supply Chain Risk (Gledhill et al., 2013)	The report examines the impact of climate change on agricultural commodities (wheat, maize and rice), energy (petroleum and gas) and mining (metal ores). The report considers the magnitude of climate change impacts and the concentration of suppliers.	In considering supply chain disruptions, the report looks at the following events and their effect on supply chains: – Heat wave and drought in Russia. – Drought in the US. – Flooding in Thailand and Australia.	The report puts forward recommendations to minimize supply chain disruptions: to not view risks in isolation, to use scenario planning, to include risk management procedures, to collaborate, and to pay attention to international resource security.	Region-specific information. The report considers the impacts of climate change over the next 10 years.
Climate Change Impacts and Risk Management: A Guide for Businesses and Government (Department of the Environment and Heritage, Australian Greenhouse Office, 2013)	Lists risks related to climate change impacts, prioritizing risks that require further attention and establishing a process for ensuring that higher priority risks are managed effectively.	Covers more frequent heat waves and hot days, the possibility of more frequent drought in Southern/Eastern Australia, sea level rise, and the increase in frequency/intensity of weather extremes (cyclones, storms/storm surges, flooding).	– Climate change risk management information/framework. – Details on how to conduct an initial assessment and a detailed analysis into the risks/uncertainties of climate change. – List of recommendations.	Country-wide (specific to Australia). The planning horizon is for 2030.
Climate Ready/UK Climate Projections (UKCP09) (accessed on 22 September 2014, available at: http://ukclimateprojections. metoffice.gov.uk/ 21678)	UKCP09 seeks to provide support for businesses in various sectors (infrastructure, built environment, health, agriculture and forestry, and local government).	Based on the UK Climate Projections that cover temperature, precipitation, air pressure, cloud, humidity. Also includes projections of sea level rise, storm surge, sea surface/sub-surface temperature, salinity, currents and waves.	The UKCP09 website also contains a weather generator which is a tool that can be used to generate statistically plausible daily and hourly time series which comprise a set of climate variables at a 5km resolution.	Region-specific (focused on UK). The projections are until the end of the twenty-first century and are relative to a 1961–1990 baseline.

Source	Description	Climate variables covered	Other information	Scale
Climate-Adapt European Climate Adaptation Platform Map Viewer (accessed on 22 September 2014, available at: http://climate-adapt.eea.europa.eu/map-viewer)	Map viewer shows observations and projections of climate change impacts, vulnerability and risks across agriculture, biodiversity, forestry, coastal areas, financial health, infrastructure, fisheries and water management.	Climate impacts/variables considered on map: – extreme temperatures, – water scarcity, – flooding, – sea level rise, – droughts, – storms, – ice and snow.	The map viewer allows the user to either select a single climate impact or show a set of maps with many climate variables considered.	Region-specific information (Europe). The time periods considered are from 2021–50 and 2071–2100.
Country Specific Model for Intertemporal Climate Version 2 (COSMIC2) (UNFCCC, 2013f)	COSMIC2 is a computer program that provides access to climate change scenarios. The user chooses one of 14 general circulation models, the country, one of 28 emission scenarios and various climate model parameters, along with a timeframe.	Country-level climate change estimates for 158 countries Climate variables considered include: – sea level rise, – precipitation, – mean temperature, – annual global mean temperature change, and – CO_2 concentration.	Key outputs include monthly mean temperature and precipitation, together with annual global mean temperature change, sea level rise, and equivalent CO_2 concentration. Most useful for estimating country-level changes.	Country-level analysis across 158 countries. The time periods considered is from 2000 to 2200.
GEO-5 for Business – Impacts of a Changing Environment on the Corporate Sector (United Nations Environment Programme, 2013)	GEO-5 assesses operational, market, reputational and policy implications of environmental trends on ten business sectors: building/construction, chemicals, electric power, extractives, finance, food/beverage, healthcare, ICT, tourism, and transportation.	Outlines key implications of environmental trends for businesses, focusing on the ten business sectors.	The document separately considers key environmental trends in detail (that is, changes in atmosphere, land, water, biodiversity, chemicals, and waste) and then considers sector specific implications from the three perspectives of operational, market and policy.	No specific time period stated, but projections appear to be until 2050. Information considered on global scale, but discusses implications for businesses.
International Threats and Opportunities of Climate change for the UK (PwC, 2013)	The report contains a thematic mapping of international threats and opportunities across business, trade and investment, infrastructure, food, health, and foreign policy.	Covers climate impacts across Africa, Asia, Australia/NZ, Europe, and America.	Contains case studies on trade, food supply chains and the insurance industry, as well as recommendations to deal with the threats and opportunities presented by climate change.	2020s, 2050s, 2080s; information available on a regional/country level.

IPCC Reports (IPCC, 2012, 2013b, 2014c) The IPCC provides the most comprehensive assessment of the available evidence on climate change. Offers some discussion of climate impacts on firms and industries.	See Chapter 2, this book.	Reports include various maps, graphs and tables in order to comprehensively detail the effect of climate change across many variables.	Timeframe ranges from now until the end of the twenty-first century.
MAGICC/ SCENGEN (Wigley et al., 2000) MAGICC/SCENGEN is a software package used in past IPCC reports to produce projections of future global mean temperature, and sea level change (corresponds to AR4).	Climate variables delivered as output in MAGICC model: – global mean temperature, – sea level change. (SCENGEN is a regionalization algorithm.)	Uses a scaling method to produce climate and climate change information on a 5 degree latitude by 5 degree longitude grid.	Regional analysis, 1990 and 2100 are the default start and end output years.
Maplecroft Climate Change Vulnerability Index (CCVI) Vulnerability index assesses 193 countries and highlights hotspots of risk.	Index evaluates exposure of countries to climate-related natural hazards, sensitivity of populations in terms of concentration, development, agricultural dependency and conflict.	The index can assist businesses to locate and monitor hotspots of risk in their operations, supply chains and locations within a specific country.	Tool can be used on regional scale, but also can be used by businesses to monitor, and manage the potential risks.
PESETA (Projection of Economic impacts of climate change in Sectors of the European Union based on bottom–up analysis) Multi-sectoral assessment of the impacts of climate change in Europe for the 2071–2100 time horizon.	The PESETA Project is based on the evaluation of scenarios based on the IPCC Special Report on Emissions Scenarios (SRES) (Nakicenovic and Swart, 2000).	Four market impact categories are considered (agriculture, river floods, coastal systems, and tourism) and one non-market category (human health).	Covers all EU countries (with the exception of Luxemburg, Malta, and Cyprus). The timeframe is until 2100.
PRECIS (Providing Regional Climates for Impacts Studies) (PRECIS, 2004) PRECIS generates regional climate projections for any region in the world with minimum area of 5000km by 5000km.	Some of the climate variables considered within the model are rainfall, precipitation, and temperature.	PRECIS allows the user to decide on the region, the Emissions scenario, the time period and required output.	Regional analysis, covers 1860–2100.
SimCLIM (SimCLIM, 2013) SimCLIM is a software package that simulates impacts of climate variations and change on sectors such as agriculture, health, coasts or water resources.	Variables considered include sea level, climate variability and extremes. Supports the representative concentration pathway scenarios from the IPCC (2013).	Contains tools for importing and analyzing both spatial and time series data.	Can be used on regional and local scales.

Source	Description	Climate variables covered	Other information	Scale
Statistical DownScaling Model (SDM) (accessed on 22 September 2014, available at: http:// www.sdsm.org.uk)	Software package produces high-resolution climate information; provides scenarios of daily surface weather variables under current and future regional climate forcing.	Climate variables covered include temperature, precipitation and humidity.	SDSM also produces a range of statistical parameters such as variances, frequencies of extremes and spell lengths.	Can be used on a regional or local scale.
The European Environment: State and Outlook 2010 (European Environment Agency, 2010)	The report includes a section on 'Adapting to Climate Change Impacts' that covers a brief assessment of vulnerability in agriculture/forestry, energy, tourism, built environment and health.	Covers key vulnerabilities related to inland water (glaciers, river floods), droughts, water scarcity, and impacts on coastal zones (sea-level rise, storms and storm surges).	The overall report includes further information on the state of the environment and also discusses risks of 'tipping elements' (climate events that have a very low or unknown likelihood of occurrence).	Historical impacts until 2009/2010; future climate change impacts until 2100; region-specific (focused on Europe).
The World Bank Group – Climate Change Knowledge Portal (accessed on 22 September 2014, available at: http:// sdwebx.worldbank.org/ climateportal/)	Portal contains maps and charts that allow user to select a region and country of interest to retrieve climate information. Within each location, there are three tabs (climate, impacts, vulnerability) showing different data for the location.	Historical climate (temperature/ rainfall), future climate and weather extremes.	Includes information on: agricultural crop projections to 2050 and contribution of agriculture sector to country's economy; fire density, flood frequency, tropical cyclone footprints; mortality risk (from floods, earthquake, cyclone, drought, multi-hazard), effects of mass disasters; and socio-economic indicators at the country level.	Global scale, time period until 2099.
Turn Down the Heat: Climate Extremes, Regional Impacts and the Case for Resilience (World Bank, 2013b)	Focuses on risks of climate change to development in Sub-Saharan Africa, South East Asia and South Asia. Report covers a range of sectors (agriculture, water, coastal fisheries). Extends the earlier 2012 report.	Climate variables considered include heat extremes, drought, aridity, water availability; also focuses on ecosystems shifts, health and poverty.	Analyses climate impacts across the main climate variables in Sub-Saharan Africa, South East Asia and South Asia; also examines the physical and biophysical impacts under different climate change scenarios.	Time periods considered are 2040s and 2080s. Regional focus on three areas: Sub-Saharan Africa, South East Asia and South Asia.

Source				
US National Climate Assessment (US Global Change Research Program, 2013)	Provides information about observed climate changes, the current status of the climate and anticipated future trends. Covers the water sector, energy supply, transportation, agriculture, forestry, ecosystems, human health, urban systems and infrastructure.	Climate variables considered include extreme weather events (heat waves, coastal flooding/storm surges, droughts), sea level rise, water availability, rising temperatures and the shrinking of glaciers.	The report also considers climate change impacts on tribal, indigenous land and native lands and resources, land-use and land cover change, and rural communities. Next national climate assessment will be released in 2014. The assessments provide input to Federal science priorities.	The report analyses current conditions, and also includes projections for the next 25 to 100 years. Region-specific (focused on US).
UNFCCC Climate Change: Impacts Vulnerabilities and Adaptation in Developing Countries (UNFCCC, 2007a)	The report offers assessments of impacts and vulnerabilities (across Africa, Asia, Latin America and small island developing states) and includes a discussion of adaptation strategies.	Covers regional information for Africa, Asia, Latin America and small island developing states on temperature, precipitation, snow and ice, extreme events and water availability.	Also discusses agriculture and food security, health, terrestrial ecosystems and coastal zones.	2020s, 2050s and 2100; regional analysis focusing on Africa, Asia, Latin America and small island developing states.
World Metrological Organization (World Meteorological Organization, 2014)	The WMO site offers a large number of information sources on climate change, impacts, covering the water, agriculture, health, energy tourism, building, transportation and humanitarian sectors.	Wide range of information on variables such as temperature, precipitation, droughts, snow and ice cover, tropical cyclones, ocean circulation and sea level rise.	The portal also provides links to policy documents, country risk and adaptation profiles, and risk management information, including the WMO Disaster Risk Reduction Programme and the Hyogo Framework for Action 2005–15.	Global scale. For future projections the user can choose a time period of 2046–65 or 2081–2100.
WRI – 'Making Climate Your Business' (World Resources Institute, 2009)	The report seeks to assist businesses (operating in Southeast Asia) to understand climate risks and the need to adapt to climate change. Focuses on five industry sectors: tourism, insurance, manufacturing, agriculture, and energy.	The report covers the impacts of weather extremes, sea level rise, reduced water supply/drought, and disease on South East Asia.	The report contains case studies to illustrate business activities that can contribute to adaptation, considers operational, considering regulatory/legal, reputational, market and financing risks.	Region-specific information (South East Asia).

Notes: A comprehensive summary of different information sources and methods is also available via the compendium on methods and tools to evaluate impacts of, and vulnerability and adaptation to, climate change (UNFCCC Secretariat, 2005).

analysis for understanding an organization's competitive position within its supplier networks. However, the impacts of climate change are generally characterized by a greater uncertainty compared to more foreseeable trends and developments in the organizational environment as well as their impacts on organizations that are easier to assess using these traditional tools for analysis. Some of these traditional analytical tools are difficult to apply under greater levels of future uncertainty given that they are static and often do not reflect trends over time or more complex interactions (Courtney et al., 1997).

Tools for evaluating adaptation needs and priorities

In order to assist businesses with evaluating vulnerabilities and the resulting adaptation needs and priorities, a number of business tools have been developed, which also include supply chain risks from climate change. These tools are meant to provide structured assessments of how vulnerable an organization is to climate impacts. The UK Climate Impacts Programme (UKCIP), for instance, has developed the Adaptation Wizard program to guide organizations through a vulnerability assessment, including an identification of current and future vulnerabilities, and to evaluate adaptation options. Part of UKCIP are resources such as 'BACLIAT' (Business and Climate Assessment Tool), a workshop-based process to assess adaptation needs and priorities. BACLIAT was originally developed with UK businesses to consider the potential impacts of future climate change. Users can also use the Local Climate Impacts Profile (LCLIP) which is a tool designed to help organizations to assess their exposure to the weather.

BACLIAT presents a generic framework for considering climate impacts on business areas, including:

1. *Markets*: What are the impacts of climate change on the demand for goods and services such as building design, technology, and global impact on markets or supply chains?
2. *Logistics*: What supply chain implications are arising from transport disruptions and impacts on suppliers?
3. *Process*: How does climate change impact production processes and service delivery?
4. *Finance*: What are the insurance and investment implications?
5. *People*: What are the impacts for employees and customers?
6. *Premises*: What are the impacts of weather on building fabric and structure as well as the internal environment?

Both BACLIAT and LCLIP can be used as stand-alone tools or as a step in a risk-based framework, such as the Adaptation Wizard.

A different approach, called the 'Climate Sensitivity Matrix', was developed by the European Commission as a guideline for project managers to assess the climate sensitivity (or vulnerability) of investments. This approach evaluates the climate sensitivity of project in relation to a range of climate variables and secondary effects/climate-related hazards. The steps involved are:

1. The sensitivity of the project options to key climate variables and hazards is systematically assessed across four key themes encompassing the main components of a value chain as follows:
    *   on-site assets and processes;
    *   inputs (water, energy, others);
    *   outputs (products, markets, customer demand); and
    *   transport links.
2. A score of 'high', 'medium' or 'low' is assigned for each project type, and theme across each climate variable.

The matrix can be useful in a business context to determine the sensitivity of project options to climate variables in relation to each of the four themes. The important climate variables and related hazards are those that are deemed of high or medium sensitivity across at least one of the four sensitivity themes.

Future scenario planning

Another approach for organizations to prepare for future uncertainty is future scenario planning. Most organizational decisions fall across a spectrum of future uncertainty, ranging from a foreseeable (or clear enough future) to true ambiguity. Some decisions face very low levels of uncertainty. Decision-makers facing Level 1 uncertainties (see Table 5.2) can develop a single forecast of the future as a basis for strategic decision-making. Fast food chain McDonald's, for example, generally faces this type of low uncertainty when deciding on the location of a new franchise business in its US market. The company has significant experience in this market, and can gather data on social and economic factors such as customer demographics, proximity to roads or transportation hubs, traffic patterns, access to its supply network, as well as other factors such as the extent of existing industry competition in a given location. While such information is often not perfect, it allows a reasonably good forecast of future restaurant earnings, and findings tend to be predictable enough to allow making a yes-or-no decision on any new US-based restaurant location (Courtney, 2003).

Table 5.2   Levels of uncertainties for decision-makers

Uncertainty Level	1 A clear enough future	2 Alternative futures	3 A range of futures	4 True ambiguity
Examples	Opening of a new outlet in a stable market.	Possible introduction of an environmental policy.	Climate change, radical innovations.	Major technological, economic, social or environmental discontinuities, abrupt climate change.
Definition	A single forecast of the future is possible and is adequate for strategy development.	A few discrete and alternative outcomes are possible, prediction relates to these.	A range of potential futures can be identified but there are no discrete scenarios. Prediction is only possible in terms of a range.	Environment is virtually impossible to identify and forecast.
Analytical tools	Traditional tools (for example, competitor analysis; Porter's Five Forces, supply chain analysis).	Decision Analysis, Valuation Models; Game Theory.	Flexible modelling/ projections, scenario analysis.	'What-if scenarios'.

Source: Adapted from Courtney et al. (1997).

Decision-makers facing Level 2 uncertainty are confronted with limited set of possible future outcomes, and the best strategy for their organization depends on which outcome ultimately occurs (Courtney et al., 1997). Typically, this type of uncertainty is associated with regulatory or legislative changes. For instance, the Australian Government introduced a package of bills to repeal the carbon tax as its first item of business in the new Parliament (see Chapter 4). At the time, it was uncertain whether or not the legislation would pass and how quickly any new arrangements, including transition arrangements, would be implemented. This uncertainty mattered to companies since the policy outcome was going to have divergent effects on companies' actions in high polluting industrial sectors. No amount of additional analysis would have facilitated a more comprehensive prediction of the outcome, even though it was possible to make assumptions on how the Senate was going to vote on this issue based on its composition (the carbon tax was eventually repealed). Organizations facing these types of situations can employ tools such as Decision Analysis (using statistical tools, such as decision tree analysis or probabilistic forecasting) or Game Theory (gauging responses to different actions, and then establishing an appropriate strategy) (Courtney et al., 1997).

The impacts of climate change fall within higher levels of uncertainty. At Level 3, a range of possible future outcomes can be identified for an organization. The range is defined by a number of key variables (for example, emissions and population growth, changes in GHG concentrations in the atmosphere), and the actual outcome may lie somewhere along the continuum within the range. Organizations conducting research on the impacts of climate change on their operations might identify a broad range of possible climate change outcomes (for instance, that the global mean surface temperature change for the time period 2016–2035 relative to 1986–2005 will likely be in the range of 0.3°C to 0.7°C, see IPPC (2013a)). Organizations would likely pursue more aggressive adaptation strategies if they knew for certain that temperature increases would be at the upper end of that spectrum. In addition to the uncertainties surrounding climate change impacts, organizations also face uncertainties in their decision-making environment due to possible changes in regulation and the number of stakeholders involved, including policymakers, communities, suppliers and buyers, to name a few. Level 3 uncertainties are not limited to climate change impacts, of course – companies in fields driven by technological innovation, for instance, are facing very similar issues in regards to evaluating their investment strategies (Courtney et al., 1997).

There are no simple answers for organizations regarding how to strategically respond to this type of uncertainty, however, scenario planning offers one option. Organizations can use scenario planning to develop a limited number of representative scenarios that offer a distinct picture of the future. Some organizations have chosen to develop 'best' and 'worst' case scenarios to represent the extreme end of the spectrum of possible outcomes, while others rely on more probable outcomes (Courtney, 2003). The example of UK water company Anglian Water (see case study below) shows how a utilities provider used climate change projections to understand the range of future impacts. The reports and methodologies used are accessible from the company's website.[2]

---

[2] Accessed 23 September 2014, available at http://www.anglianwater.co.uk/ environment/climate-change/adaption.aspx.

CASE STUDY: USING CLIMATE DATA TO ASSESS IMPACTS AND VULNERABILITIES – ANGLIAN WATER

Anglian Water is a water company operating in the east of England. Anglian Water sources its water from a low lying region which has a long coastline, thus making the area susceptible to climate change impacts. This has prompted managers at Anglian Water to take proactive adaptation measures.

In 2005 and 2010, Anglian Water used the UKCP02 and UKCP09 climate change projections to conduct company-wide risk assessments (the UKCP02 is the older version of the UKCP09 projections, see Table 5.1 for details). These risk assessments considered the vulnerability of abstraction, treatment and supply assets. In addition, the company's business unit specialists worked with organizations such as the UK Meteorological Office, the Tyndall° Centre (Manchester University), UK Water Industry Research (UKWIR), and the UK Environment Agency (EA) in order to develop specific assessments for impacts on key business activities (UNFCCC, 2014).

In 2010, Anglian Water created a Climate Change Steering Group to better manage risks and action plans for adaptation and mitigation. Anglian Water also appointed a full-time climate advisor who has had the role of coordinating education and research in order to embed climate change adaptation into the company's operations (Anglian Water, 2014).

CASE STUDY: DESIGNING TO WEATHER, CLIMATE AND CLIMATE CHANGE – RIO TINTO

Mining company Rio Tinto initiated its interest in adaptation measures in 2002 when it undertook an internal climate change risk assessment. The company's first adaptation study involved considering the actual impacts of weather extremes and projected climate changes that were outlined in the IPCC *Third Assessment Report*. Rio Tinto subsequently followed this with a second study that focused on the consequences of climate change in more detail. Rio Tinto conducted this study by asking the UK's Hadley Centre for Climate Change to provide a summary of future climate change scenarios, showing how climate variables might change over the next 50 years. The purpose of this analysis was to allow the company to better understand vulnerabilities in the geographic regions where Rio Tinto has significant mining interests (Nitkin et al., 2009).

These studies allowed Rio Tinto to conclude that upgrades to existing structures might not be necessary over the short- to medium-term. However, the study showed that Rio Tinto's exposure to climate change risk would vary by location in the long-term. Following these studies, Rio Tinto conducted comprehensive site assessments of its higher priority areas. The purpose of such assessments is to gauge changes in potential changes in cyclonic activity and topographic effects (Nitkin et al., 2009). Rio Tinto's main business concerns regarding the impacts of climate change relate to the availability of water – particularly risks of flooding (too much water) or droughts (too little water). In response to concerns regarding this risk, Rio Tinto has developed a water strategy to facilitate the effective management of the risks of both future droughts and floods (Nitkin et al., 2009).

Level 4 uncertainty is characterized by true ambiguity and does not provide decision-makers with any steady basis or prior experience to assess or forecast the future. Uncertainties are unknown and also unknowable. Such situations are quite rare, and are likely to occur when facing major technological, economic, social or environmental discontinuities (Courtney, 2003). In the context of climate change, Level 4 uncertainties arise from the possibility that climate surprises (Streets and Glantz, 2000) or abrupt climate changes may occur. Although there is a large body of research on the social and ecological impacts of climate change, much of this research has relied on scenarios with slow and gradual changes. In part, this reflects that scenarios that are regarded as 'less likely' or 'not likely' have not gained widespread recognition, and that it has been difficult to generate appropriate scenarios of abrupt climate change for impacts assessments (Alley et al., 2003).

Despite such challenges, there have been attempts at 'imagining the unthinkable'. For instance, Schwartz and Randall (2003) have assessed the consequences of an abrupt climate change scenario to better understand the potential implications on the United States' national security. In their research they have applied a hypothetical 'what-if' scenario (what if an abrupt climate change event would occur) and have reviewed several iterations of the scenario with leading climate change scientists. Their report outlines that such a scenario would lead to food shortages due to decreases in net global agricultural production, decreased availability and quality of fresh water across key regions due to changes in rainfall patterns, more frequent floods and droughts, as well as disruptions in access to energy supplies, with wide-spread implications for geopolitical stability. The report has not remained without criticism, in part because of the underlying presumptions leading to the creation of the hypothetical scenario situation. Nonetheless, the report points to the importance of assessing a wider range of scenarios and the importance of identifying existing vulnerabilities that could be exacerbated under more extreme outcomes.

Concluding Comments

This chapter has looked at two main challenges arising from climate change for organizations: first, climate change impacts, and second, vulnerabilities as they affect the corporate sector and create challenges to adapt and the need to create resilience. We have reviewed traditional and emerging drivers for organizations to respond to a changing environment and put forward frameworks for evaluating both impacts and vulnerabilities. Taken together with the significant uncertainties surrounding the impacts of climate change, it is evident that much remains to be learned about 'what actually works' when it comes to adaptation (McGray, 2013), and that successful adaptation will not be easy to achieve (Linnenluecke, 2013a). A great challenge related to adaptation is that outcomes are difficult to assess due to cost-benefit asymmetries (that is, near-term investments are not leading to near-term outcomes), and the consequences of the types of impacts that an organization cannot adapt to easily (which will thus need resilience capacities). We will now further discuss these issues in the next chapter.

6.   Strategic options for adaptation

The previous chapter introduced the concept of a 'coping range' as a framework that can be used for understanding the relationship between a changing climate and organizational adaptation. This concept allows organizational decision-makers and stakeholders (such as employees, suppliers, buyers, creditors or investors) to develop an understanding of which impacts an organization can and cannot cope with and which impacts will lead to vulnerabilities. This information can subsequently be developed into quantitative information (Carter et al., 2007; Jones and Boer, 2005). The concept of a coping range can also be expanded to assess current and future adaptation options and their outcomes, as well as planning and policy horizons (Willows and Connell, 2003; Yohe and Tol, 2002). Nonetheless, adaptation options and successes are difficult to evaluate – whether or not adaptation has positive outcomes can vary based on actors and the timeframes involved. Difficulties also arise as costs are usually relatively well known or assessable – while adaptation benefits are dependent on climate change outcomes and impacts, which are difficult to project. This chapter focuses on strategic actions for adaptation that organizations can undertake to broaden their coping range and details emerging methodologies to evaluate the costs and benefits of implementing these options.

In response to evaluating adaptation options, a variety of approaches and frameworks have been developed which have improved over time. Older assessments, also referred to as 'first generation' or 'type 1' assessment studies (Burton et al., 2002), are often top-down assessments. These top-down assessments are based on scenarios that are downscaled from global climate models to the local scale through several analytical steps, beginning with global data and then moving through biophysical impacts towards a socio-economic assessment (Carter et al., 2007). The 'second generation' assessment studies (Burton et al., 2002) pay greater attention to information around vulnerability to inform decisions on adaptation. They also consider the involvement of stakeholders in decision-making around adaptation options (Füssel and Klein, 2006; LDC Expert Group, 2012). This chapter continues to build on the frameworks introduced in the previous chapter that include 'top-down' assessments of impacts and 'bottom-up' assessments of vulnerability, including the assessment of both biophysical climate change impacts and the factors that make organizations vulnerable to those impacts.

Decision-making for Adaptation

The previous chapter has introduced the assessments of impacts, focusing on the biophysical climate change impacts to which organizations and industries need to adapt, as well as the assessment of vulnerability, focusing more specifically on the risks arising from climate change and the propensity to be harmed (Adger, 2006). This section now focuses on the identification and evaluation of strategic adaptation options. The identification of adaptation options should thereby be based on the assessments of impacts of vulnerability, to examine the adaptive capacity, and adaptation measures which are required to broaden the coping range of an organization exposed to climate change (Noble et al., 2014; Smit and Wandel, 2006). The generic elements of assessing adaptation options are summarized here, and align with frameworks proposed in our earlier research and by the UK Climate Impacts Programme (UKCIP) for the assessment of climate change.

1.　**Identifying exposure:** Based on assessment of impacts and vulnerabilities (see Chapter 5), this step includes an assessment of climate change impacts as they affect the region(s) or location(s) in which the organization is operating, also considering changes in policy, economy, society or technology that could exacerbate or mitigate climate change impacts. Furthermore, this step involves an assessment of flow-on effects from climate change impacts that could affect an organization's supplier, buyer, or resource base, and that could disrupt organizational activities in addition to direct impacts, and an assessment of how vulnerable the organization is to these changes based on the coping range model. Sieve mapping (or overlay/hazard mapping) can provide a visualization of climate impacts by displaying data graphically on a base map that includes information about organizational or industry-relevant locations.

2.　**Identifying priority areas:** In many cases, not all parts of an organization face the same level of exposure to climate change, and some parts may be less vulnerable than others. This step requires an assessment of the level of risk of disruption through climate change impacts at different organizational locations and along its supply chain, and the identification of high-risk or priority areas for adaptation.

3. **Identifying adaptation objectives:** This step requires the definition of a particular objective(s) the organization is seeking to achieve through adaptation, which could include reducing vulnerabilities or minimizing supply chain disruptions. The adaptation objectives should be related to specific timeframes (for example, short- or long-term goals) and specific parts or aspects of the organization that have been identified as being particularly exposed and at risk of disruption from climate change.

4. **Identifying adaptation options:** Adaptation options broadly fall within three categories outlined in the previous chapter: *structural/ physical options* (improved maintenance and management of infrastructure, strengthening of infrastructure foundations, protection of critical assets, enhancing redundancies, and – where needed – considering relocation), *social options* (raising awareness and education, improving information), and *institutional options* (incentives through taxes or subsidies, insurance, zoning laws and building standards, national and regional adaptation plans, and improved disaster planning and forecasting) (Noble et al., 2014). An organization should identify a range of adaptation options in a first step, and then evaluate their costs and benefits in achieving a certain objective.

5. **Establishing decision-making criteria and appraising options:** In order to achieve a specific adaptation objective, an organization should decide on the criteria for evaluating different adaptation options. The evaluation of adaptation options is increasingly moving away from a simple cost-benefit analysis and the identification of 'best practice' examples to the development of multi-metric evaluations including risk and uncertainty dimensions in order to provide more sophisticated support to decision-makers (Chambwera et al., 2014). Decision-criteria can include not just costs in relation to benefits, but also a range of other outcomes.

6. **Making decisions:** Any choice of adaptation strategy is likely to involve some trade-offs and can include barriers to adaptation, including costs, behavioral biases and limited resources available to the organization (Chambwera et al., 2014).

7. **Implementing decisions:** The implementation of any strategy or decision often brings its very own set of issues. These include the need to convince internal and external stakeholders about the necessity of the decision, but also issues around the actual implemented option that may not quite resemble the intended option, as a range of stakeholders are typically involved in decision-making. Consequently, the management of the implementation process and

the subsequent monitoring, evaluation and review are important steps to feed back into organizational learning and adjust any maladaptive outcomes.

8. **Monitoring, evaluating and review:** Adaptation is not a static issue and evolves over time in response to both a changing climate (changing impacts) and changing vulnerability (Hallegatte, 2009). Consequently, adaptation decisions should be monitored and reviewed via a long-term and flexible process that allows for learning and adjustment. Generally, the literature indicates that optimal adaptation and the desirability of particular strategies will vary over time depending on the level of climate change and other factors such as the availability of different technological adaptation options and their maturity (Chambwera et al., 2014).

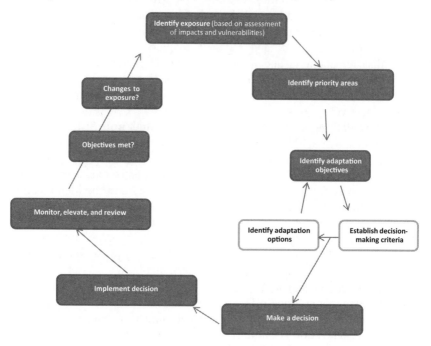

Source: Adapted from UKCIP (2004).

Figure 6.1   Assessing options for adaptation

A range of similar approaches have been identified by other authors. Hintz et al. (2011) suggest an approach that comprises the following steps: first, identifying the most relevant hazard; second, estimating the total climate risk – this figure should include expected losses based on current climate patterns and projections of future risk; third, identifying what can be done to minimize these risks using infrastructure-based responses, new technologies, new processes, behavioral change, insurance or other financial means of transfer; fourth, applying a thorough cost-benefit analysis (CBA) to the measures identified; and fifth, creating a prioritized portfolio based not only on costs and benefits, but also on a range of other considerations, such as the ease of implementing measures from a political perspective.

The Climate Adaptation Tool (Bellamy and Aron, 2010) suggests three assessment steps. First is identifying and prioritizing climate risks drawing upon the UKCP09 climate projections to inform climate risk identification. This can be replaced with projections for other regions. Risk prioritization is achieved by assigning quantitative scores to the likelihood and severity of both climate vulnerabilities and opportunities for organizations. Second is identifying and appraising adaptation options – this step provides a multi-criteria appraisal methodology to assess the performance of different options (an optional CBA feature is integrated). Third is auditing the performance of implemented options – this step provides a framework to assess the performance of implemented adaptation options against multiple criteria.

The framework in Figure 6.1 provides an overview of the key contingent variables in an adaptation assessment. At times, adaptation is implemented reactively as demonstrated by Entergy's response to the US hurricanes Katrina and Rita (see case study below). However, decisions to adapt are often not based on rigorous assessment steps. First, adaptation decisions are shaped by the social and economic contexts in which an organization operates, often leading to adaptation constraints. Changes in the economic climate, policy decisions, technological changes or societal pressures can either inhibit or support climate adaptation decisions. Furthermore, adaptation is guided by decision-makers who are contemporaneously 'sense-making' of climate change patterns and subsequent choices and actions and dependent on their assessment of the level of risk of disruption to operations through climate change impacts at the organization's current location (Linnenluecke and Griffiths, 2011). We discuss these issues in further detail below.

CASE STUDY: CLIMATE CHANGE ACTION IN RESPONSE TO A NATURAL DISASTER – ENTERGY

In 2005, over the Gulf Coast of the United States, hurricanes Katrina and Rita caused major loss of life and impacted organizations in the infrastructure sectors (transport, energy), tourism, and petrochemical sectors, resulting in global effects on oil prices and insurance losses of US$60 billion and US$15 billion, respectively (Huppert and Sparks, 2006; Munich Re, 2007, 2009).

New Orleans-based energy provider Entergy was prompted to act in response to climate change following its US$2 billion loss as a result of damage caused by hurricanes Katrina and Rita. Following the hurricanes, Entergy took immediate action to strengthen its emergency response performance by relocating its data and transmission centers to areas that were less prone to flooding and storms, and by setting up a business continuity group to evaluate the consequences and impacts of climate change and other business threats to its activities.

The business continuity group conducted a three-phase adaptation analysis (Sussman and Freed, 2008). The first phase of the analysis involved a study identifying climate and risk drivers over the short-term (20 years), medium-term (20–50 years) and long-term (until the end of the twenty-first century). The analysis thereby considered a number of climate variables such as droughts, earthquakes, temperature changes and sea level rise, and used trend analysis and historical probability distributions to develop scenarios for these key variables. The second phase involved examining the correlation of each identified risk from the first phase with Entergy's assets or operations. The third phase involves assessing existing risk mitigation plans and formulating alternative plans to reduce the negative impacts of climate change (Sussman and Freed, 2008).

Assessing barriers to adaptation

As part of identifying adaptation options, organizations should also focus on identifying adaptation barriers (or adaptation constraints). Such barriers should be identified in adaptation decisions so that they can be factored in, and so that possible alternative options can be discussed. In some areas, public sector engagement will be required through policy developments, direct public investment, tax measures, or national or international institutions for adaptation coordination (Chambwera et al., 2014). Such measures should ideally be developed in conjunction with industry and community consultation to maximize outcomes across scales, and to avoid maladaptation.

Cost barriers to adaptation include costs associated with making and implementation adaptation decisions. For instance, organizations may incur significant costs for acquiring information about climate change and assessing their vulnerability that is not readily available (Kelly et al.,

2005). These costs may include expenses for educating staff, obtaining climate and weather data, or the need to invest in building additional capabilities and conducting additional research. In many cases, organizations are faced with the question if they should update or replace existing infrastructure and carry the adjustments costs. These costs can be a significant barrier to early capital replacement in infrastructure-dependent industries with large-scale assets (Fankhauser et al., 1999). In addition, some socially desirable actions may not be cost-effective for organizations. For example, organizations may choose to not invest in flood mitigation measures when flood risks are partly assumed by insurance or post-disaster support (Burby et al., 1991; Laffont, 1995). Organizations may find that incentives to adapt are lacking, for instance in areas such as biodiversity protection and the conservation of natural assets (Chambwera et al., 2014). Many organizations also still face limitations around assessing costs and benefits of adaptation, which we will further discuss in this chapter.

CASE STUDY: BARRIERS TO ADAPTATION – ANGLIAN WATER

In its 2011 *Adaptation Report*, Anglian Water identified four main barriers to adaptation implementation:

1. absence of clear national guidance;
2. lack of nationally-coordinated research;
3. variability in customer acceptance of the need for adaptation; and
4. costs of adaptation not are recognized in price limits and there are difficulties in valuing loss of service (Anglian Water, 2011).

Policy barriers can arise if governments are not sufficiently supporting adaptation, or are not removing adaptation barriers. Governments face similar issues as private organizations when evaluating impacts and vulnerabilities. There are often limited financial resources, and significant levels of bureaucracy to be overcome, possibly leading to coordination failures within the government, especially in cases where multi-ministry actions are required. For example, the reduction of flood risks may require coordination among a country's environmental ministry and the ministry of finance, as well as local authorities (Jha et al., 2013). Government action is at times also driven by a focus on narrow interest groups (Chambwera et al., 2014).

Externalities can occur if adaptation actions by one organization, industry or even country create higher damages for others. One example is transboundary water supply where increased irrigation in one country can create water scarcity downstream (Goulden et al., 2009). Trans-sector effects can also take place, for instance, when adaptation in one sector creates the need to adapt in another sector (for example, see Attavanich et al., 2013).

Lastly, the *uncertainty around future climate change outcomes* is a significant barrier to successful adaptation. Decisions about adaptation actions have to be made without fully knowing future impacts and vulnerabilities (Chambwera et al., 2014).

Improving sensemaking

Organizational decision-makers constantly seek to adapt to external conditions, however, decision-makers do not always make full use of available information in supporting their decision-making, especially long-term projections of consequences (Camerer and Kunreuther, 1989; Henriet and Michel-Kerjan, 2008; IPCC, 2014a; Thaler, 1999). For instance, while there is already information available on possible adaptation options, adaptations decisions by organizations and industries are often not following optimal recommendations. Sensemaking theories developed in the social sciences seek to explain how corporate decision-makers arrive at an understanding of the organization's environment, and strategic actions that need to be priorities. These theories show how issues are being interpreted by decision-makers, and then subsequently translated into action. Sensemaking has been defined as 'turning circumstances into a situation that is comprehended explicitly in words and that serves as a springboard into action', and as 'efforts at sensemaking tend to occur when the current state of the world is perceived to be different from the expected state of the world' (Weick et al., 2005).

Climate change is often not receiving much attention in boardrooms as it is perceived to be a low-salience risk issue. Gradual warming changes cannot be directly experienced by individuals, meaning that it is difficult for decision-makers to 'sense' that change is happening (Whiteman and Cooper, 2011). Vulnerability to already observed changes (for example, already observed temperature changes) is often known and can be assessed retrospectively. However, the key challenge lies in understanding vulnerability to future impacts which requires not only an understanding of future climate change impacts, but also of organizational, industrial and other developments. Cognitive and sensemaking limitations are often not easy to overcome. To better understand and 'make sense' of climate change, a first step for organizations is an engagement with the science around climate change, and also with the aforementioned assessment frameworks – such a proactive engagement is an important first step in identifying impacts, vulnerability and exposure.

Identifying Adaptation Options

Any organization interested in adaptation should consider a variety of adaptation options. There are no distinct recommendations in regards to which adaptation is the 'best' adaptation option – this is dependent on the sector and the particular challenges an organization is facing. No-regret, low-regret and win-win strategies have proven particularly successful, given that they do not expose an organization to a great level of risk when making the 'wrong' decision. Win-win strategies are strategies that allow achieving multiple objectives, such as adaptation and mitigation, or economic, environmental and social sustainability aspects (Hahn et al., 2010). However, incremental adaptive responses may not be entirely sufficient, given the continuing uncertainty about the extent of climate change (Noble et al., 2014). Generally, a number of options are open to organizations. Organizations can adapt to actual or anticipated changes, which requires speed, flexibility and agility in creating a portfolio of options (for example, by diversifying across products, regions or suppliers that vary in their exposure to climate change). Organizations can also take a leadership position and set new standards or create new demand for products and services that facilitate adaptation (Courtney et al., 1997). In this chapter, we review two options in greater detail: *In situ* adaptation, and geographical diversification (or relocation). Organizations should consider their advantages and disadvantages given the variant projected levels of climate impacts.

In situ **adaptation options**
Table 6.1, below, lists a range of possible adaptation options in line with recent suggestions by the Intergovernmental Panel on Climate Change (IPCC) (Noble et al., 2014). Some of these measures fall outside of the scope of an organization, nonetheless, the case of Eskom shows how organizations can become actively involved in shaping adaptation outcomes. Several of the options are also overlapping, and can thus be pursued simultaneously. In addition to the information presented here, internet portals such as the weADAPT platform by the Stockholm Environmental Institute (SEI) and partners (weADAPT.org) enable decision-makers to access information, share experiences and insights into 'lessons-learned'. The platform includes tools for knowledge integration, an 'adaptation layer' – a Google Earth interface to show who is

doing what, a climate adaptation knowledge base, customized user and organization profiles, and adaptation decision support tools (such as tools to screen and prioritize adaptation options).

Table 6.1 Adaptation options and examples

Adaptation Option	Examples
Structural/physical	
Adaptation of engineered and built structures	Changes to building codes, integrating climate risks and sea level rise. Protection measures against climate impacts and extreme weather events (seawalls, flood levees, coastal protection, drainage, updates to water and electricity grids). Infrastructure adaptation (enhanced maintenance, strengthening, protection, enhanced redundancies).
Technological	New crop and animal varieties, genetic options. Conservation agriculture. New technologies (irrigation, water saving, rainwater harvesting, cooling and ventilation, renewable energy) and technology transfer.
Ecosystem-based	Maintaining healthy ecosystems, ecological restoration and conservation, including wetlands, floodplains and mangrove forests. Afforestation and reforestation, maintaining biodiversity. Green infrastructure (for example, trees for shade) and ecological corridors. Assisted migration or managed translocation. Community-based natural resource management. Adaptive land-use management and management of fisheries.
Safety nets	Seed and food banks. Public health, sanitation, social safety nets and social protection (primarily developing countries).
Social	
Educational/ Informational	Education and awareness about risks, including media communication. Sharing local and traditional knowledge, integrative adaptation planning. Knowledge-sharing platforms and research networks.
Improved data sets and warning systems	Climate services (forecasts, downscaling, longitudinal datasets). Hazard and vulnerability mapping, systematic monitoring and remote sensing, early warning and response systems. Participatory scenario development, including traditional or indigenous knowledge.

Adaptation Option	Examples
Behavioral	Community preparedness. Changing practices in agriculture, water management and other sectors. Income diversification.
Institutional	
Financial	Financial adaptation incentives (subsidies, taxes). Insurance and incentives from insurers to avoid maladaptation. Catastrophe bonds. Payments for ecosystem services. Disaster contingency funds. Microfinance.
Laws and Regulations	Changes in land zooming laws, building standards and building codes. Water regulations and agreements. Disaster risk reduction initiatives, including encouraged insurance purchasing. Creation of protected areas (marine areas, fishing).
Policies and Programs	Adaptation plans (national, regional and local levels). Upgrading of critical infrastructure (energy and water management, health, communications). Disaster planning and preparedness. Landscape, watershed and integrated coastal zone management. Sustainable management of areas such as forests, fisheries. Community-based adaptation.

Source: Adapted from Noble et al. (2014).

Several studies show the importance of adapting buildings and engineered infrastructure to rising sea levels and other climate-related risks (Hamin and Gurran, 2009). However, many infrastructure developments (water supply networks or transportation infrastructures) are still based on historical climate analysis, meaning that novel risks are not factored into new developments. Since infrastructures are built for long lifespans, adaptations in new building projects can facilitate adaptation from the onset – while retrofitting existing buildings and infrastructure is often very costly, if not impossible in the case of large-scale and embedded infrastructures (for example, road and transportation networks). Challenges certainly arise, especially as there are still uncertainties around future climate projections. The 2013–14 drought in California shows the importance of adaptation in national water policy frameworks and supply infrastructures to cope with climate extremes.

CASE STUDY: ADAPTING ROAD INFRASTRUCTURE TO
CLIMATE CHANGE – EGIS

Climate events have a notable direct impact on road infrastructure, and can
cause loss and damage on both an economic and physical scale. While in the
past infrastructure has been designed on the basis that the climate in the future
will not be substantially different to that of today, many organizations have now
recognized that this is a dangerous presumption.

In 2012, Egis, a French engineering group involved in the areas of infrastructure and transport systems, planning, water and environment, produced a report
titled *Adapting Road Infrastructure to Climate Change: Innovative Approaches
and Tools.* This report presented a number of adaption options for road
infrastructure to climate change risks including:

● Assessing transport sector preparedness to climate change.
● Assessing hydro-meteorological issues when designing transport infrastruc-
  ture.
● Identifying transport infrastructure vulnerabilities to climate change.
● Forecasting and managing extreme climate events in real-time.
● Alerting infrastructure owners, operators and users.
● Managing climate change risks for transport infrastructure and networks
  (Ennesser 2012).

The solutions presented in the report were based upon research undertaken at
both a national and international level (UNFCCC, 2014).

Organizational relocation as adaptation

More extreme adaptation measures such as relocation have been considered in scientific and policy-oriented discussions. However, relocation as
a strategy has not yet received much attention in discussions on organizational responses to climate change. Limits to *in situ* adaptation are
predominantly evident for organizations in industries such as agriculture,
herding or fishing that are highly dependent on ecosystems (Linnenluecke and Griffiths, 2011; Warner et al., 2009). If natural resources are
negatively affected by climate change, the production base of sectors
dependent on these resources will be directly and adversely affected.
From a climate change perspective, a main question is whether organizations and industries in highly-affected regions and localities can
actually undergo successful *in situ* adaptation (see above), or whether it
would be a better economic decision in the long-run to *relocate*. There
are likely to be push- and pull-factors that influence relocation decisions
(Arauzo-Carod et al., 2010; Brouwer et al., 2004; Pallenbarg et al.,

2002). Such decisions may be motivated by the desire to find a better location for an organization's operations (for instance, a site which is less vulnerable to climate impacts), but also by decisions to move away from the present location (Linnenluecke and Griffiths, 2011).

Organizations and industries are highly embedded in their existing supply chains and industry networks as well as their political, economic, social and technological environment (Hess, 2004; Romo and Schwartz, 1995; Uzzi, 1996). Consequently, a strategy of relocation will be – in many cases – not a first choice when it comes to adaptation. Any decision to invest in new products or facilities in a new location or region (irrespective of whether or not climate change is an influencing factor) is determined by a range of issues, of which climate change is just one. Factors that are likely to play a role in organization location and relation decisions include production costs, the sophistication of markets and infrastructure, legislative and taxation burdens, subsidies, and access to natural and human resources, as well as access to finance and other essential organization inputs (Linnenluecke and Griffiths, 2011).

The framework in Figure 6.2 (adapted from Linnenluecke and Griffiths, 2011) provides an overview of key factors influencing organization relocation decisions based on the impacts of climate change and an organization's vulnerability. The impacts of climate change can include both direct impacts from changing temperature and weather patterns, but also indirect impacts from flow-on effects, such as effects from weather extremes affecting the reliability of infrastructures and supply chains (Porter and Reinhardt, 2007). Decision-makers need to carefully evaluate the trade-offs between relocating into regions possibly less affected by climate change, associated costs which could be high (especially for resource and infrastructure-dependent industries). Relocation costs and benefits are thereby dependent on location-bound advantages. Constraining factors for relocation decisions are high levels of organizational embeddedness within a specific location or industry, high costs associated with search, and high levels of uncertainty regarding the relocation outcomes (Brouwer, 2010; Knoben and Oerlemans, 2008; Pallenbarg et al., 2002). If organizations face high levels of uncertainty and high levels of relocation costs, they are unlikely to consider relocation (Linnenluecke and Griffiths, 2011). Furthermore, institutional factors such as government support, subsidies, and economic, political or technological factors are likely to play a role in facilitating or hindering organizational relocation (Linnenluecke and Griffiths, 2011).

If a relocation is not a desirable or feasible option, decision-makers can consider relocating or adapting organizational infrastructure or activities locally (that is, *in situ*, see section above), especially if the organization

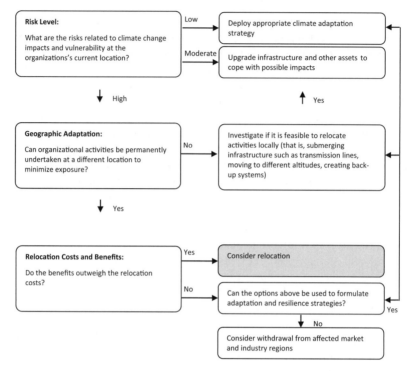

Source: Adapted from Linnenluecke and Griffiths (2011).

Figure 6.2   Relocation decisions

faces significant local vulnerabilities to the impacts of climate change. Examples include erecting sea-level walls, upgrading infrastructure foundations, moving to different elevations (a strategy used in alpine tourism to improve snow reliability), or temporary relocation options such as back-up facilities to increase organizational-level resilience against the unexpected (Linnenluecke and Griffiths, 2010). The importance of back-up systems and infrastructure becomes evident when looking at the experience of energy company Entergy during hurricanes Katrina and Rita (see case study above). The company had to respond to the highest number of customer outages in its history, which resulted in expenditures of about US$1.5 billion to finance the restoration efforts and forced its New Orleans-based subsidiary to file for bankruptcy protection (Vince et al., 2007). In addition, the company had to temporarily relocate staff and management functions to other cities. The availability of back-up management systems or infrastructure might have greatly helped to facilitate recovery.

Establishing Decision-Making Criteria

Many adaptation measures such as different building standards, sea walls or different irrigation techniques will require substantial upfront investment. A common assumption is that well-targeted, early investment to adapt to the impacts of climate change will likely be cheaper and more effective in the long run, compared to not adapting and dealing with impacts as they arise. However, there are large associated uncertainties in regards to the nature and impact of future climate change outcomes and possible adaptation costs and benefits, posing impediments to decisive action on adaptation. Adaptation outcomes are also often reliant on actions taken by others (for example, by legislators or other organizations along the supply chain), thus adding additional levels of uncertainty. This chapter looks at the important questions of how to evaluate adaptation options in terms of their benefits. In contrast to mitigation efforts, that is, efforts targeted towards the reduction of greenhouse gas (GHG) emissions, there are no established frameworks to evaluate the effectiveness of different adaptation options across organizational and industry levels over time. The cost and benefits of mitigation strategies can be established through mechanisms for evaluating mitigation efforts such as GHG emissions accounting. Mitigation strategies have been linked to financial performance benefits for organizations as emission reductions efforts encourage resource savings (for example, energy) and thus decrease expenditure for resource inputs. Even though organizational-level mitigation efforts are in many cases not linked to global emission targets, organizations can evaluate their progress and benchmark themselves against others by tracing their carbon footprint.

While methodologies for evaluating mitigation have become more established, the assessment of adaptation needs, costs and the benefits associated has proven to be more complicated. The *IPCC Fourth Assessment Report* (AR4) reported that the literature on understanding adaptation costs and benefits has been 'quite limited and fragmented' (Adger et al., 2007) and concluded that 'many adaptations can be implemented at low cost, but comprehensive estimates of adaptation costs and benefits are currently lacking' (Parry et al., 2007: 69). Similarly, studies by the Organization for Economic Co-operation and Development (OECD) on adaptation costs and benefits (Agrawala and Fankhauser, 2008) concluded that little quantified information is available to understand the link between investment into adaptation and possible benefits and outcomes. Since the release of the AR4 and the OECD study, some noteworthy reports on a policy and on country-level contexts were undertaken to fill the existing gaps (for instance, by the Economics of Climate Adaptation Working Group (2009)). In terms of organizational and industry-level adaptation efforts, however, there exist no official indicators to track progress on adaptation, and the relationships (or trade-offs) between costs and benefits are not well understood. Survey data shows that few organizations have established comprehensive adaptation objectives with corresponding strategies, plans and activities, as well as business indicators to track adaptation progress (United Nations Environment Programme, 2012).

CASE STUDY: INTERNATIONAL UNION OF RAILWAYS

The International Union of Railways (UIC), which is the worldwide international organization of the railway sector, developed the Adaptation of Railway Infrastructure to Climate Change (ARISCC) Project in 2009 (International Union of Railways, 2012). The project was developed for the purpose of facilitating the implementation of adaptation strategies and measures for transport systems and railways (Nolte et al., 2011).[1]

The main outcomes and objectives of the project included the identification of the main impacts of climate change on railway infrastructure, the assessment of risks and costs of climate change, and the development a guidance document for railway infrastructure managers. The guidance document was developed to assist managers in identifying vulnerable assets, choosing adaptation measures and assessing the costs and benefits of adaptation (Nolte et al., 2011). The UIC undertook the project to allow managers to effectively deal with climate-related hazards such as flooding and severe storms, and to avoid or minimize damage to railway infrastructure assets. The process began with an assessment of how to manage natural hazards under today's weather and climate conditions, and then involved the development of strategies to deal with future weather and climate conditions.

The ARISCC Project draws on two case studies. These included the railway lines in the Rhine Valley from the Alpine regions at the source to the flatlands at the estuary, and the West Coast Main Line running from London to Glasgow in the UK. These case studies were selected for investigating and assessing adaptation options against a defined set of criteria, and for comparing them with scenarios where no adaptation takes place. Case studies are also used within the project to identify the most vulnerable locations and assets.

The final outcomes are available via the ARISCC website, accessed 23 September 2014, http://www.ariscc.org/.

Organizations can draw upon a wide range of tools and techniques for evaluating and prioritizing adaptation options. These include qualitative

---

[1] Within the overview of the ARISCC Project it is acknowledged that the challenge for railways includes not only adaptation, but also resilience planning to address extreme weather events, and to recover speedily from any resulting impacts.

assessments (such as stakeholder consultations, discussions with focus groups and experts, or ranking exercises), scenario-planning exercises or quantitate modeling approaches that usually require an analysis of costs and benefits, as detailed below. In this chapter, we predominantly focus on two different approaches: the cost-benefit analysis (CBA) and the multi-criteria analysis (MCA). The CBA is a commonly used tool in decision-making. However, the evaluation of adaptation options is increasingly moving away from just looking at cost-benefits towards more sophisticated, multi-metric evaluations. These include risk and uncertainty dimensions in order to provide more sophisticated support to decision-makers (Chambwera et al., 2014). Organizations can certainly also appraise their own approaches to risks, that is, whether they are happy to accept certain risks or require adaptation measures (Brown et al., 2011). In any analysis the ancillary effects should also be considered – these can be positive or negative outcomes of adaptation. For instance, the upgrading of building standards and the inclusion of insulation to protecting against heat may mitigate energy use and associated GHG emissions, but may also lead to higher building costs and/or more material usage (Chambwera et al., 2014; Sartori and Hestnes, 2007).

Adaptation costs and benefits (cost-benefit analysis)
A key issue in undertaking any adaptation-related CBA is to understand the benefits of adaptation. While adaptation costs (for instance, different building standards) can generally be estimated using market prices and discount rates, an estimate of benefits is often not readily available. Difficulties arise because adaptation focuses on 'moving targets' such as changing temperature and sea levels. The benefits of adaptation will vary significantly depending on the level of environmental change. Additional problems arise from factoring in unexpected or unpredictable outcomes of climate change, less-than-optimal decision-making by individual decision-makers, as well as the costs associated with overcoming barriers to adaptation (such as the preference for, or aversion against certain adaptation options held by decision-makers and local communities). In the absence of detailed organizational-level estimates of adaptation needs, costs and benefits, we broadly canvas alternative approaches for quantifying the economic costs and benefits of a wide range of adaptation measures. Decision-makers seeking to evaluate their organization's vulnerability and exposure to climate change along with adaptation needs should keep in mind that climate change impacts will vary among different locations. Climate risks will be location-specific, and so will be impacts on an organization's activities, as well as the cost and benefits associated with any adaptation decisions.

Table 6.2   Selected approaches for analyzing adaptation costs and benefits

Approach	Approach Summary	Organizational Context Application
Cost Benefit Analysis (CBA) (*United Nations Framework Convention on Climate Change, 2011*)	CBA can be used to assess adaptation options when efficiency is the only decision-making criteria. *Steps involved:* 1 Determining adaptation objectives and potential adaptation options. 2 Establishing a baseline (situation without adaptation intervention being carried out). 3 Quantifying and aggregating costs of an adaptation option over a specific time period (direct and indirect). 4 Quantifying and aggregating benefits of an adaptation option over a specific time period (direct and indirect). 5 Comparing aggregated costs and benefits.	*There are three ways to use CBA:* 1 Net present value (NPV) approach to assess difference between present value of benefits and present value of costs. 2 Benefit-cost ratio approach to look at ratio of present value of benefits to present value of costs. 3 Internal rate of return (IRR) approach: discount rate at which the NPV equals zero. *Potential difficulties:* – Challenging to include reliable estimates of criteria that are valuable to organizations, but not valued in markets (for example, social/cultural values). – Difficult to measure potential ancillary benefits of climate change adaptation. – CBA does not assess equity considerations related to the distribution of costs and the benefits of adaptation options across stakeholder groups. – Difficulties with choosing appropriate discount rate.

Cost Effectiveness Analysis (CEA) (*United Nations Framework Convention on Climate Change, 2011*)	Approach assesses costs of adaptation options. *Steps involved:* 1 Determining adaptation objectives and potential adaptation options. 2 Establishing a baseline (namely, the situation without adaptation intervention being carried out). 3 Quantifying and aggregating various costs. 4 Determining the effectiveness – the definition of effectiveness depends on the adaptation objectives and the established baseline. 5 Comparing the cost-effectiveness of the different options.	*Application:* – Used to find the least costly adaptation option or options for meeting selected physical targets. – Applied in assessing adaptation options in areas where adaptation benefits are difficult to express in monetary terms (for example, human health). *Potential difficulties:* – CEA is difficult to use as sole decision-making tool because the benefits are defined in a single dimension only (namely, cost-effectiveness).
Costs of Adaptation (Focus on Water Sector) (*Parry et al., 2009*)	Approach to estimating future adaptation or residual damage costs at water management or scheme level. *Steps involved – simple approach:* 1 Estimating (over the next few decades) capital and operating costs (over all actors) plus residual damages, in the absence of climate change (but including other changes, such as the effects of changes in demand or exposure) 2 Repeating calculations, assuming a climate change trend and a management strategy specifically designed to cope with this trend. *Steps involved – advanced approach:* 1 Assuming a specific adaptation decision is made and then estimating costs and residual impacts under a range of possible future climates. 2 Assigning likelihoods to possible climate outcomes, then estimating expected costs and impacts under the defined adaptation action. 3 Repeating this analysis with alternative adaptation decisions and identifying a range of possible costs and residual impacts.	*Application:* Applicable to management level, considers locally specific adaptations that vary with geographical, financial, institutional and socio-economic circumstances. *Potential difficulties:* – Assumes knowledge about factors.

Approach	Approach Summary	Organizational Context Application
Economic Framework (Department of Climate Change and Energy Efficiency, 2012)	Approach is used to assess costs and benefits of adaptation options. *Steps involved:* 1 Scoping the study to set project parameters. 2 Quantifying cost impacts of historical weather events. 3 Quantifying changes in future weather events – determining the number and magnitude of future weather events. 4 Modeling impacts without adaptation. 5 Modeling impacts with adaptation – determining the NPV of adaptation options. 6 Analyzing portfolios of adaptation and communicating the findings.	*Application:* Framework can be used in an organizational setting. *Potential difficulties:* Difficult to forecast ancillary benefits or benefits to organizations that are difficult to value.
Implementation Analysis UNEP (Feenstra et al., 1998: 135)	Method focuses on identifying the least costly adaptation measure. User must identify any implementation barrier and then evaluate how difficult or easy it will be to overcome this barrier. Matrix can be used to identify barriers, actions to overcome the barriers, what time and financial resources are required and the degree of difficulty in overcoming the barriers .	*Application:* Tool can be used for any region or location. The approach is useful when it can be assumed that the benefits of different adaptation measures will be comparable. *Potential difficulties:* Difficult to analyze non-comparable adaptation measures.
Marginal Abatement Cost Curve (MACC) (McKinsey and Company, 2008)	MACC is an accounting methodology used to present graphically the investment performance of different energy water and waste reduction projects. Methodology ranks various projects from the most cost-effective to the least cost-effective while illustrating the total carbon, water or waste abated by each individual project.	*Application:* Approach is used to identify adaptation costs. This approach can be used in any context that looks at different adaptation projects. This approach can be used in an organizational setting to assess and analyze the cost-effectiveness of adaptation options.

UKCIP Costing UK Climate Impacts (UKCIP, 2004)	The UKCIP costing methodology assists users in calculating the costs of climate impacts and describes how to compare these to the costs of adaptation measures. *Steps involved:* 1 Identifying and measuring climate impacts in physical units. 2 Converting the physical impacts into monetary values. 3 Calculating the resources costs of the proposed adaptation options. 4 Weighing up the costs and benefits of the adaptation options.	*Application:* Useful for determining adaptation costs and needs. The spreadsheet tool has been specifically developed for use within the UK, but the costing method can be used anywhere.

Multi-criteria analysis

Multi-criteria analysis (MCA) assessments provide opportunities for establishing more comprehensive decision-making criteria to evaluate adaptation options beyond a financial assessment. In addition to a cost analysis (including total costs and cost effectiveness), organizations can also use one or several of the following criteria (Brown et al., 2011; Carter and Mäkinen, 2011; Smit et al., 1999):

- Effectiveness in the short-, medium-, and long-term and under different scenarios of future climate.
- The time required to establish a certain option in relation to the urgency of adapting to certain impacts.
- The lifetime or duration of an adaptation option.
- The flexibility and reversibility of a particular option versus the desire to commit.
- The possible 'level of regret' when making the wrong decision.
- Secondary or cross-sectoral impacts, externalities or co-benefits.
- Opportunity costs.
- Effectiveness at supporting resilience and reducing the impacts of extreme events (for example, floods, droughts, strong winds).
- Social and environmental implications.
- Preferences from a broad set of stakeholders.
- Limiting factors for implementation or sustainability (for example, resource constraints).
- Other outcomes, such as opportunities for learning or building capacity (Smit et al., 1999).

Table 6.3 summarizes a number of approaches to MCA. Table 6.3 is by no means exhaustive, but illustrates some possible avenues for conducting such assessments. The choice of criteria ultimately depends on the adaptation project, as well as the objectives the organization is seeking to achieve. Typically, a broader range of assessments is likely to lead to more detailed insights.

Table 6.3   Multi-criteria analysis

Approach	Summary of approach	Organizational Context Application
Multi-criteria analysis (MCA) (United Nations Framework Convention on Climate Change, 2011)	*Steps involved:* 1 Determining adaptation objectives and potential adaptation options. 2 Agreeing on decision criteria – criteria need to be described including unit and span of possible scores. 3 Scoring performance of each adaptation option against each of the criteria. 4 Assigning a weight to criteria to reflect priorities. 5 Ranking the options – the total score for each option is calculated by multiplying standardized scores with their appropriate weight.	This approach is useful where cultural and ecological considerations are difficult to quantify and when monetary benefit or effectiveness are only two of many criteria. *Application:* – Can incorporate both quantitative and qualitative information. – Allows for stakeholder engagement by allowing beneficiaries of adaptation options to be involved in choosing criteria and options. *Potential difficulties:* – Assignment of weights can be difficult if the number of criteria is very large. – Possible disagreement on the weight assigned to each criteria.
Screening of Adaptation Options (Mizina et al., 1999)	All adaptation options are entered into a matrix to allow for a screening according to pre-defined evaluation criteria (typically entered across the top of the matrix). These can include aspects such as: 1 Does the adaptation measure have to be implemented in advance? 2 Is a decision being made now that could incorporate climate change (window of opportunity)? 3 Will the adaptation option generate other benefits (for example, economic or environmental)? 4 What are the costs? 5 What is the feasibility? The end-user can insert, or substitute other criteria, if they deem such other criteria more appropriate.	*Application:* Tool is best used at beginning of decision-making process, allowing the user to create a more manageable list of adaptation options. Can be used on an organizational level.

Approach	Summary of approach	Organizational Context Application
Tool for Environmental Assessment and Management (TEAM) (Julius and Scheraga, 2000)	Software package that creates graphs and tables allowing users to compare the relative strengths of adaptation strategies using both quantitative and qualitative criteria. Assists user in evaluating issues such as equity, flexibility and policy coordination. User lists strategies across the top of the table and the evaluation criteria down the side, and then enters a score indicating the relative performance of each strategy under the various criteria. Table can then be used to construct a variety of graphs derived from the core data.	*Application:* Useful for evaluating adaptation options. Tool can be used for any region or location.
Adger et al. (2005)	Adger et al. (2005) introduce multiple criteria against which adaptation can be judged: 1 Effectiveness: the capacity of the adaptation option to achieve its expressed objectives. Two key indicators of the effectiveness of an adaptation option are robustness to uncertainty and flexibility. 2 Efficiency: assessment of efficiency requires consideration of: (i) the distribution of costs and benefits; (ii) costs and benefits of changes in goods that cannot be expressed in market values; (iii) timing of adaptation options. 3 Equity and legitimacy in adaptation: adaptation can be evaluated from the perspective of outcome (that is, who wins and who loses from the action) as well as who decides on the adaptation to take. Legitimacy is determined by looking at the extent to which decisions are acceptable to participants and non-participants that are affected by those decisions.	The authors introduce these criteria as evaluative criteria for judging adaptation success at different scales.

Concluding Comments

This chapter has focused on the strategic actions for adaptation that organizations can undertake to broaden their coping range, and has also detailed the emerging methodologies to evaluate the costs and benefits of implementing these options. The chapter has thereby built on the frameworks introduced in the previous chapter which include 'top-down' assessments of impacts and 'bottom-up' assessments of impacts, including the assessment of both biophysical climate change impacts and the factors that make businesses and other organizations vulnerable to those impacts. It has proposed a framework for the identification and evaluation of strategic adaptation options based on eight steps:

1. Identify exposure;
2. identify the priority areas;
3. identify the adaptation objectives;
4. identify the adaptation options (including *in situ* adaptation and relocation);
5. establish the decision-making criteria and appraise the options;
6. make the decision;
7. implement the decision; and
8. monitor, evaluate and conduct a post-implementation review.

This chapter has also looked at the challenges that adaptation outcomes produce, namely difficulties in assessment due to cost-benefit asymmetries (that is, near-term investments are not leading to near-term outcomes), and the consequences of the types of impacts that an organization cannot adapt to easily (which will thus require resilience capacities).

In the next chapter, we will now look in more detail at those impacts that the organization cannot adapt to – or indeed those impacts that fall outside the coping range, even of the most well-adapted organizations.

7. Responses to extreme environmental changes

Gradual, longer-term environmental changes associated with climate change are most likely the types of changes to which organizations can more easily adapt to. Of particular concern not just for organizations and industries, but also for society at large, are changes in the number and/or frequency of weather extremes (such as heavy rainfall, heat waves or storms) and potentially even more drastic and lasting changes in the Earth's climate (Alley et al., 2003; Wilbanks et al., 2007b). Any climate-change related changes to the frequency and/or severity of extreme weather events will be particularly impactful for sectors with a high risk of exposure – those dependent on large-scale infrastructure including the energy, transportation and agriculture sectors, as well as those managing, financing or insuring these infrastructures (Linnenluecke et al., 2012). However, while extreme events are particularly impactful, gradual changes may also build up and exceed thresholds at which effects become notable (Wilbanks et al., 2007b). For instance, the extensive drought in California in 2013–14 has posed significant strains on California's system of dams and reservoirs, thus reaching critically low levels of water supply. The State Water Project, which provides added water supplies to a large percentage of the state's population and farmland, had to reduce allocations.

It is certainly important to note that not all changes in climate will automatically result in negative consequences for all industry sectors. Stern (2007), for instance, has concluded that Russian agriculture might benefit from a warming climate to some extent by being able to grow crops over longer periods and in regions that used to be unsuitable for farming. However, this does not mean that the risks of climate change should be underestimated. A large and growing number of research studies point to significant adverse impacts for most sectors and locations, especially in relation to the impacts of extreme weather events (IPCC, 2012). While most organizations are usually equipped to deal with some form of variability in their natural environments such as seasonal changes, the impacts of climate change pose novel risks that fall outside previous experience and can exceed coping ranges, even of those organizations that have established adaptation measures. Any changes that occur more abruptly, are more persistent and/or have impacts that are large may lead to vulnerabilities for organizations, industries and society, especially if these impacts not anticipated and/or if there are no or limited opportunities for adaptation (Schneider et al., 2001; Wilbanks et al., 2007b). Consequently, organizations with a narrow coping range might not be able to withstand impacts. Furthermore, impacts might not just be limited to the organization as such, but may also extend to its supply chains, resource access or distribution channels, and the access to customers and markets (Linnenluecke and Griffiths, 2011).

Limits to Adaptation? Responding to Impacts Exceeding the Coping Range

Dow et al. (2013), in their recent paper in *Nature Climate Change*, raised the important question of whether there are limits to adaptation, and what the consequences of reaching such limits might be. Limits to adaptation can be defined as 'obstacles that tend to be absolute in a real sense: they constitute thresholds beyond which existing activities, land uses, eco-systems [...] cannot be maintained, not even in a modified fashion' (Moser and Ekstrom, 2010). Dow et al. (2013) conclude that attempts have been made to identify adaptation limits for ecological systems (for example, species extinctions), but that little is known about adaptation limits in human systems. This finding certainly translates to organizations and industries – little is known in regards to whether there are organizational or industry-specific limits to adaptation, what the factors are that define or influence them, as well as what the consequences of reaching or exceeding such limits might be.

Limits to adaptation become particularly visible during exposure to more extreme conditions which exceed the coping range. Going back to the example from Cyclone Larry in the opening paragraph of this book, the force of such a cyclone easily exceeds measures to deal with weather extremes. As a result, several industries suffered major losses in gross value of production. Particularly affected was the banana industry (A$283 million) as most banana producers in the affected region lost their entire crop. Due to the length of the banana growing cycle, this meant little or no income for affected producers for nine months until new crops could be harvested. Other affected industries included: the sugar industry (A$111 million in raw sugar output, also due to areas of lost crop); the fruit and nut trees industry ($58 million) where many producers suffered extensive damage to trees and almost total loss of crops, with restoration efforts taking approximately 5–15 years; the forestry (\$5 million), vegetable (\$4 million) and lifestyle horticulture (\$1 million) industries even though the key production areas are not centered in this region; and the livestock industries (\$4 million) due to livestock losses from wind and flooding (Department of Primary Industries and Fisheries, 2006). Additional impacts were felt in a number of industries such as tourism, insurance, transportation, to name a few.

However, research has also demonstrated that less 'visible' limits to adaptation exist, such as ecologically defined limits related to the inability of crops, animals or humans to deal with heat stress (Sherwood and Huber, 2010). Humans have, in many cases, been able to stretch or overcome physical and other ecological limits with innovation, such as breeding programs or genetic modifications of crops to increase heat resistance (Moser and Ekstrom, 2010). Due to such human intervention, limits to adaptation have been difficult to establish or define. However, given that the impacts of climate change and weather extremes will become increasingly impactful, opportunities to adapt may be finite for many actors, ranging from individual households to businesses or governments (IPCC, 2012). Cases of maladaptation due to policies that increase vulnerabilities to climate change, unintended consequences of adaptation actions, as well as unintended consequences of human interventions in ecosystems have already been documented (Bradshaw et al., 2004; Haines et al., 2006). Furthermore, different actors are likely to hold different risk evaluations and experiences, and thus differ in their willingness to undertake adaptation action (Dow et al., 2013).

Depending on future climate change outcomes, key regions such as low-lying coastal areas might ultimately become unavailable or unsuitable for economic activity. Organizations may face what is implied by the 'population ecology' perspective (Hannan and Freeman, 1977) – that organizations unable to sufficiently adapt to a changing environment are selected out. Similar issues may also arise for organizations unable to shift from a high carbon to a low carbon future. Exxon Mobil, for instance, has come under increased pressure after publishing a report stating that the company does expect increasing government action to curb GHG emissions, but that such action would not reach the level required to limit temperature increases to below the 2°C target – implying a continued reliance on carbon extraction (ExxonMobil, 2014; see also Chapter 5 on stranded assets). The following sections look at different strategic options for organizations to cope with the more extreme environmental changes resulting from climate change.

Accepting losses

One option for organizations is to accept the risks and associated possible losses from the impacts of climate change and weather extremes. This option involves both bearing losses and sharing losses. Loss bearing is a decision typically left to the individual organization. However, it can also be pursued in circumstances where there are no other choices (for example, in cases where insurance is not available or too costly for the organization to take out). Losses can be shared within wider communities, or via mechanisms such as public relief (De Loë et al., 2001). The greatest concern about accepting losses is that the resilience of organizations may erode over time if they are exposed to repeated impacts from climate change.

Building organizational resilience

Resilience is a characteristic of organizations that possess a sufficiently wide coping range and/or can quickly recover from situations that create vulnerability to their operations once the boundaries of the coping range have been exceeded (Linnenluecke and Griffiths, 2012). Organizations can therefore either attempt to broaden their coping range through adaptation, or respond to impacts exceeding the coping range by taking actions and implementing strategies to cope with impacts and losses that are unavoidable. The notion that organizations have, or are able to develop resilience has largely been inferred from research findings in other fields such as engineering, material sciences, ecology or psychology which show that engineered systems, materials, ecosystems or individuals which possess certain characteristics can handle adversity in a way that allows them to recover from disturbance, disruption, or change, and at times even in a strengthened fashion (Linnenluecke and Griffiths, 2013; Woods, 2006). Based on these findings, researchers have suggested that organizations with resilience capacities (that is, certain organizational structures, resources or processes) can recover or even emerge stronger when confronted with adversity. These claims were supported by observations that some organizations were more successful in surviving the impacts from financial market shocks, terrorism attacks or industrial accidents than other organizations under similar circumstances (Gittell et al., 2006; Linnenluecke and Griffiths, 2013). Resilience is, thereby, not constrained to post-disaster or post-impact recovery, but also includes the idea of preparing for unknown risks and consequences – as far as possible – and 'improving' of essential basic structures and functions (IPCC, 2012).

Resilience has increasingly become of interest in the popular press and academic research. Resilient organizations are generally seen as those organizations that can better deal with abrupt, unexpected change or adverse impacts in comparison to others. However, despite the growing utilization of the concept in the popular press and academic research, there are still few insights into the conceptualization and operationalization of the concept. A number of commentators (for example, Klein et al., 2003; Manyena, 2006) have argued that it is necessary to have a solid understanding of, first, the factors determining resilience in a certain context (that is, the types of internal capacities or external support that allow an organization to absorb impacts and recover), and second, how organizations can create and improve resilience.

The resilience literature tends to examine not just individual organizations and their resilience, but focuses on interconnected actors and how they are embedded within larger systems (Allenby and Fink, 2005). Organizations can be understood as multi-interacting actors, embedded in complex social, ecological and geophysical systems (Linnenluecke and Griffiths, 2013). Individual organizational resilience is thereby still important – to ensure that the individual organization is not prone to failure and can cope with unforeseen impacts and surprises. However, the larger social, ecological and geophysical systems can greatly influence organizations and entire sectors – and vice versa. For instance, organizations and industries in regions or localities that provide strong financial institutions, recovery funds, emergency responses and reconstruction support often prove more resilient in the aftermath of disasters than those organizations and industries located in areas that do not possess these features.

The recognition of the systems perspective has been driven by increasing connectivity (within and across scales) in terms of communication flows, resource flows and connections between organizations, markets and supply chains. Importantly, the nature of the systems the organization is embedded in can potentially undermine long-term resilience. For instance, organizations located in ecosystems with declining resilience due to climate change impacts may experience direct adverse impacts on their profitability, especially when not paying attention to ecosystem processes and engaging in unsustainable practices such as overfishing, overfarming or other types of resource exploitation. The resilience perspective also maintains that organizations may become highly adapted to deal with specific risks – but that they are not prepared to deal with impacts outside of those risks. For example, an agricultural organization may be well adjusted to cope with variations in temperature, but may be unprepared to cope with more hot days or increasing instances of drought.

The sections below discuss factors that have been found to contribute to resilience in prior studies. Many options for building resilience involve building excess capacity that allows organizations to become less susceptible to disaster impacts such as those from drought, floods or storms. However, for all resilience options outlined below the caveat needs to be applied that they can lead to inefficiencies transactions costs or costs associated with duplication if they are not carefully considered, implemented and managed (Fankhauser et al., 1999). The costs and benefits associated with different options for building resilience can be evaluated using the tools and frameworks discussed in previous sections, with the difference that resilience also focuses on actions and strategies to cope with impacts and losses that are unavoidable through adaptation.

Flexible infrastructure design   A significant driving factor in fostering resilience is to consider how impacts can be minimized, and how recovery can be maximized. The ex-ante integration of adaptation measures, such as carefully selecting building sites to avoid risk factors or the integration of suitable national building codes can in many cases minimize the impacts of climate change impacts by widening the coping range. However, once impacts exceed the coping range, organizations need more flexible infrastructure decisions to respond to larger-than-expected or more-severe-than-expected weather-related risks. We find such models in the pastoral (beef production) industry in Australia where pastoral organizations possess large land masses that allow a rotation of cattle in the instances of drought. Similar models that allow geographic diversification (be that within the value chain or among suppliers, buyers or target markets) can provide greater flexibility and diversified income options. Critical infrastructure (such as data centers) or relief and overflow sites can be designed in a way that they can be moved to safer locations and support organizations during relief and reconstruction phases. In addition, opportunities to decentralize an organization's workforce can not only provide resilience against the impacts from climate and weather extremes, but also provide resilience against a number of other disruptions (Allenby and Fink, 2005).

Back-up operational and technical capacity   Organizations that have structures in place that allow them to quickly mobilize back-up operational and technical capacity are often better protected against significant impacts. Similarly, the back-up of data and information can also ensure that organizational data is not wiped and at once, which can help to protect against catastrophic loss (Allenby and Fink, 2005). The importance of data back-up systems has become evident also during the 11

September 2001 attack on the World Trade Center, where organizations such as bond-trading firm Cantor Fitzgerald were able to rapidly resume operations as they had established back-up data facilities as part of their business continuity and contingency plans (Allenby and Fink, 2005).

Financial and relational reserves   Generally, organizations that have viable organizational or business models in place are more resilient that others, mostly because they have financial reserves in place that assist in protecting the core of the organization, and that also help the organization to rebuild without having to look into layoffs or other cost-cutting measures. Research undertaken on airlines' responses to the 11 September attack shows that those airlines that maintained adequate financial reserves were able to preserve relational reserves, and vice versa. These reserves contributed to the resilience of these organizations in the time of crisis (Gittell et al., 2006).

Insurance and financial aid agreements   An obvious response to rising levels of risks and impacts from climate change is to increase insurance levels. Alternative risk transfer instruments have become popular, such as index-linked securities, including catastrophe bonds and weather derivatives (IPCC, 2012). Within the agricultural sector, weather derivatives have been introduced as new financial products that derive their value from basic variables such as temperature, precipitation or wind. While insurance requires a demonstration of loss, weather derivatives require no such demonstration and pay out based upon observed weather. The claim is often made that weather derivatives and similar products balance out the financial impact of weather and weather extremes on affected organizations, and thus allow a smoother adaptation to climate change. However, this claim has not been without criticism, questions arise whether these products allow organizations to effectively insuring themselves against specific losses, or whether they are just placing bets on future weather conditions (Australian Bureau of Statistics, 2003; Bates, 2013).

Communication flows   In addition to infrastructure design, another important factor is the design of organizational and inter-organizational communication channels and appropriate warning systems to allow for effective communications and advanced warning in cases of emergency. In many cases, communication breakdowns or insufficient information and warming times have had significant adverse effects on organizational, industry and community resilience. For instance, during the 2009 Victorian bushfires in Victoria, Australia, the emergency management system experienced a breakdown due to the unprecedented amount of information flows. Resulting impacts were confusion among the population in regards to the impending fire threat, as well as significant problems in providing a sufficient level of emergency services to the local population (Linnenluecke and Griffiths, 2013). As climate change impacts are increasing, better communication flows will be required to communicate risks among a range of stakeholders.

Community support The role of community responses is increasingly considered to be of importance for creating resilience, both for minimizing the impacts of any adverse event and for responding in the aftermath. Strong local community networks fostered by cultural or social ties can support effective responses (Ford et al., 2006). The aftermath of disasters often reveal the importance of community action. In the aftermath of the 2009–10 Brisbane Floods, groups of residents, volunteer teams, local community and church groups and networks of residents came to the aid of local businesses and residents in the immediate aftermath of the flood. This 'civic engagement' (Putnam, 1993) proved highly effective in many cases – especially where local emergency management teams were overwhelmed and slow to respond. In the aftermath of Hurricane Katrina, there was a measurable disparity between local recoveries across communities, leaving many communities devastated. While there is often a high level of urgency to have communities restored as quickly as possible, long-term benefits can be gained through carefully implemented reconstruction (Hallegatte, 2008; Hallegatte and Dumas, 2009; IPCC, 2012).

Improvisation and learning Spare capacity that allows for learning and innovation is also seen as an important factor that increases capacities to address surprises or external shocks (Folke et al., 2005). The ability to learn from previous experience, the appropriate management of people, improved communication, and the willingness of organizational members to be mindful towards and address negative impacts are all seen as important factors in improving resilience. Some redundancy increases the chances that social memory is being retained and can be activated (Ostrom, 2009). Furthermore, having capacities that allow for learning and innovation also increases the chances for improvisation if the organization does experience a significant, adverse impact (IPCC, 2012). The resilience to climate change will in most cases depend on innovation, the development of new ideas and novel solutions, or the expansion of familiar ideas and options to meet emerging new needs and to respond to surprises (Denton et al., 2014).

Supply chain effects   Given that supply chains are becoming increasingly interconnected and complex as a result of global sourcing efforts, supply chain resilience has become a topic of significant importance. Vulnerabilities to climate change are no longer constrained to individual organizations, but are extending internationally. Consequently, factors external to an organization, such as disruptions in the supplier base or transportation network, can negatively impact overall resilience. The design and reliability of supply chains therefore needs to take the impacts of climate and weather extremes into consideration.

CASE STUDY: INCREASING SUPPLY CHAIN RESILIENCE – BRITISH SKY BROADCASTING (BSKYB)

BSkyB is a company engaging in the provision of entertainment, news, sport, and movie programs to over 10 million customers across the UK and Ireland. A major climate change risk facing BSkyB is that of dramatic weather patterns that hinder the company's ability to reliably source the natural resources needed for BSkyB's set-top boxes and other electronic devices (Carbon Disclosure Project, 2012). Ensuring a reliable and cost-efficient source of rare earth metals through increasing supply chain resilience is pivotal to the company.

The company realized that in order to remain competitive in the goods and services it offers, it would not only need to be energy-efficient in its operations, but also engage in successful and effective adaptation to climate change and build resilience against climate change impacts. Key to BSkyB's policy of increasing supply chain resilience is to guarantee access to key inputs. This is achieved through sourcing its products from several geographical locations, and engaging in a stock management system which ensures that the company has enough stock if flow is hindered for a period of time (Carbon Disclosure Project, 2012).

BSkyB has also engaged in active measures to minimize its dependence on inputs by implementing a process to reuse and recycle all its products returned to it through its engineers or via freepost. Doing this has allowed the company to create a dependable closed-loop supply chain (Carbon Disclosure Project, 2012). The company also has an Executive Environment Steering Group (EESG) that meets every eight weeks to identify and assess risks and opportunities, and to determine company-wide strategies (Carbon Disclosure Project, 2012).

CASE STUDY: IMPROVING CUSTOMER CONFIDENCE –
CARIBBEAN TOURISM ORGANIZATION

Tourism in the Caribbean Region has historically exhibited seasonal variability due to the region's exposure to significant climate change impacts, such as temperature extremes and increased frequency and strength of natural disasters in the summer months.

The susceptibility of the Caribbean to such climate change impacts, particularly in summer, has the potential to discourage tourists from visiting the Caribbean region in the warmer months. This variability in demand prompted the Caribbean Tourism Organization in the 1990s to engage in marketing efforts to portray the Caribbean region as a four-season destination.

This was achieved by advertising campaigns that emphasized a number of attractive and pertinent new features of tourism resorts, such as upgraded air-conditioning, new hurricane interruption policies and discounted room rates (Nitkin et al., 2009). In recent years, many companies in the Caribbean region have also introduced hurricane guarantees that provide a replacement stay of the same duration or equivalent value as the original booking (Nitkin et al., 2009).

This adaptation strategy has proven to be successful. Following the promotional campaigns, occupancy rates in many beach resorts over the summer season began to rise and are currently almost equal to occupancy rates in the winter season (Nitkin et al., 2009).

Resilience in critical infrastructure systems　　Climate change, and climate-change related changes in weather extremes, can cause great disruption to critical infrastructure systems, including electricity networks, transportation routes, water supplies, communications systems and health providers. Disruptions to critical infrastructure organizations often compounds recovery due to limited access to essential services. Consequently, planning for infrastructure resilience is vital and should take into consideration how climate change will impact critical infrastructures over many decades. Especially large-scale infrastructure is difficult to change as it exhibits strong inertia and irreversibility and thus requires large investments in change. In these instances, strategies of anticipation and planning are required to avoid vulnerabilities.

Concluding Comments

This chapter has looked at avenues for organizations to respond to climate change impacts that exceed the coping range. Different options exist for organizations to cope with such impacts. One avenue is to simply accept risks (and associated possible losses), which may erode organizational resilience over time. A second option for organizations is to consider building resilience against those impacts. The chapter discussed a number of different options, ranging from infrastructure design to back-up capacity, financial and relational reserves, insurance mechanisms as well as improved warming systems and community support mechanisms. Many options for building resilience relate to building excess capacity that allows organizations to become less susceptible to disaster impacts, such as those from drought, floods or storms. However, the costs and benefits associated with different options for building resilience need to be carefully considered and evaluated. Decision-makers should also consider the interdependence of their organization with ecosystems and social systems which can either support or hinder organizational resilience.

8. The path forward: new frameworks for business strategy and innovation

Humanity has already emitted and continues to emit a large amount of carbon into the atmosphere. The growing levels of GHG emissions already make a complete avoidance of the adverse impacts of climate change unachievable. Even the strongest attempts at stabilizing GHG concentrations would not be able to prevent all residual loss and damage, and especially vulnerable locations, communities and sectors will still feel some of the effects of climate change. Climate change, therefore, calls for new approaches that take into account not only the need to mitigate GHG emissions, but also the need to adapt and the need to build resilience (Denton et al., 2014). The role of businesses in responding to climate change is clear. While businesses have been central to the creation of the wealth and the technologies that have transformed society (Michaelis, 2003), organizational development and economic growth have been major drivers behind the intensification of natural resource use and consumption that has risen to unsustainable levels. Following assessments in earlier chapters regarding how organizational decision-makers can assess climate change impacts, vulnerabilities and the prospects for adaptation and creating resilience, this concluding chapter summarizes avenues for approaching climate change through a combination of different strategic responses. In terms of 'what to do' to respond to climate change (now and in the future), this chapter identifies and discusses the formation of climate-resilient pathways (Denton et al., 2014).

In line with the recent 2014 Intergovernmental Panel on Climate Change report, climate-resilient pathways are defined in this chapter as development trajectories that combine adaptation, mitigation and resilience strategies, and as actions that are aimed at reducing climate change and climate change impacts (Denton et al., 2014). While adaptation, mitigation and resilience are often treated as distinct concepts, they all depend on societal change and technological innovation, and thus on overall response capacity (Moss et al., 2010). Climate-resilient pathways can therefore be understood as iterative and continually evolving processes that help to achieve the fundamental objective of the United

Nations Framework Convention on Climate Change (UNFCCC), defined in Article 2 of the UNFCCC as a stabilization of GHG concentrations in the atmosphere 'at a level that would prevent dangerous anthropogenic interference with the climate system'.[1] Furthermore, the Convention text argues that: 'such a level should be achieved within a time frame sufficient to allow ecosystems to adapt naturally to climate change, to ensure that food production is not threatened and to enable economic development to proceed in a sustainable manner'.

At what level we reach 'dangerous anthropogenic interference with the climate system' is still debated. Overall, the literature agrees that there is a certain level of climate change that is low enough that most systems (such as ecosystems or social systems) could adapt without much effort. However, there is also a level of climate change that is high enough that these same systems may not be able to cope with the impacts (Denton et al., 2014; Rockström et al., 2009a,b). Between those two levels, the challenges to adapt to climate change increase as the level of climate change rises (Denton et al., 2014). How exactly these levels of climate change correspond to global average temperature increases is not fully known. Chapter 1 looked at the commonly cited threshold or 'guard rail' for 'safe' levels of temperature rise, the so-called 2°C target, as the upper limit of warming permissible to avoid dangerous anthropogenic interference in the climate (Randalls, 2010). Even though there is an unclear scientific foundation for this target, it was nonetheless adopted by the Conference of the Parties (COP) to the UNFCCC. Decisions reached in Cancun at COP-16 state that 'deep cuts in global greenhouse gas emissions are required according to science, as documented in the Fourth Assessment Report of the Intergovernmental Panel on Climate Change, with a view to reducing global greenhouse gas emissions so as to hold the increase in global average temperature below 2°C above pre-industrial levels'.

Considering recent outcomes of international climate policy negotiations, it is clear that there is a very significant gap between the aggregate emission reduction pledges by Parties in terms of global annual GHG emissions by 2020, and the aggregate emission reductions necessary to achieve a stabilization of global average temperature increase below 1.5°C or 2°C above pre-industrial levels. Since COP-16, formal recognition was given to the goal to limit the global average temperature increase to a maximum of 2°C above pre-industrial levels, with pressure

---

[1]   The full text of the Convention, accessed 23 September 2014, is available at http://unfccc.int/resource/docs/convkp/conveng.pdf.

from developing countries to consider a limit to 1.5°C (see Chapter 3). However, it remains to be seen whether a UNFCCC target of 1.5°C or 2°C is the 'right' target to avoid 'dangerous anthropogenic interference with the climate system'. As discussed in Chapter 1, GHG emissions corresponding to a specified maximum warming are poorly known. A target of 1.5°C or 2°C above pre-industrial levels does, therefore, not represent a truly reliable threshold (Meinshausen et al., 2009).

Despite the emerging consensus in international negotiations that global temperature increase needs to be limited and remain beyond a certain target, scientists are unconvinced that the consequences of reaching climate thresholds (such as a certain amount of temperature increase) are understood well enough to support any recommendations regarding specific warming targets or thresholds. Scientists have also raised concerns whether adaptation and measures to build resilience are really understood well enough to support a clear determination of the amount of adaptation and resilience required to avoid adverse impacts (Denton et al., 2014). Scientific findings point to the need for early actions, and for careful and iterative assessments of adaptation, mitigation and resilience strategies to determine their overall effectiveness. While in the past organizational and policy decision-makers did not seem to consider climate change response strategies as a top priority and critical issues to address, this situation seems to be changing (Linnenluecke, 2013b). Many decision-makers are realizing that climate change has significant and destructive impacts on businesses, industry and society.

Integrating Strategic Approaches

This chapter makes the point that an integrated strategy drawing together adaptation, mitigation and resilience is essential for climate change risk management at all scales. Delayed action in the present may reduce options for climate resilient pathways in the future. Mitigation is an important preventative strategy to keep climate change impacts moderate, rather than extreme. Adaptation and resilience are response strategies to those impacts of climate change that cannot be (or are not) avoided under different scenarios of climate change (Denton et al., 2014). While mitigation can keep climate change impacts moderate, rather than extreme, adaptation as a response strategy can keep the vulnerabilities to climate change moderate, rather than extreme (Denton et al., 2014).

The benefits and costs of integrating adaptation, mitigation and resilience will vary according to the circumstances of each particular industry sector, organization and locality (De Boer al., 2010; Wilbanks, 2003). As detailed in previous chapters, appropriate responses aimed at adaptation, mitigation and resilience also differ from situation to situation, given that different types of industries and sectors are affected differently by climate change, and that each organization or sector faces different issues on multiple scales depending on the political, economic, social, technological, and institutional context. For instance, in highly vulnerable sectors and locations, adaptation and resilience may be a higher priority, mostly because of the benefits that are to be obtained by reducing vulnerabilities to the impacts of climate change and weather extremes. Resilience, in addition to adaptation, is thereby an important risk management strategy to address the possible impacts of climate change that organizations have not (or cannot) adapt to, as these impacts exceed the coping range.

Adaptation initiatives are still seen as a low priority (especially in highly developed countries) due to perceptions that various actors within society can 'wait-and-see' and have abundant adaptive capacity once future climate impacts become visible (Næss et al., 2005). This perception is changing, mainly because of the major losses and damages that even some industrialized countries have started to experience due to climatic variability and the impacts of extremes events. For instance, the US experienced significant impacts from hurricanes, droughts and severe winter storms in recent years. Mitigation, on the other hand, is seen as a more pressing response to policy and stakeholder pressures, especially in sectors where organizations contribute to a large proportion of GHG emissions.

Irrespective of local circumstances, most sectors share at least one fundamental similarity: climate change impacts and vulnerabilities are greater if the impacts of climate change are substantial rather than moderate (Denton et al., 2014). Given the multi-level context in which organizations are embedded, effective institutional environments can thereby support and coordinate joint outcomes, and facilitate the development of climate-resilient pathways. While climate-resilient pathways often exist in distinctive local contexts, they are shaped by external linkages that connect these local contexts across geographic scales and time. For example, mitigation cannot be achieved in a few isolated places if it is not achieved in other places (Denton et al., 2014). Similarly, adaptation might not be successful if it is only undertaken by a single organization in isolation without considering issues in the supply chain, such as resource supply constraints.

When attempting to find climate resilient pathways (and win-win solutions between adaptation, mitigation and resilience), decision-makers need to consider a variety of questions. These include questions regarding the type(s) of strategies that need to be implemented to respond to the magnitude and rate of climate change, questions regarding possible trade-offs in outcomes, as well as questions how successful decision-making can occur at and across different scales (Adger et al., 2005). These points are further discussed below. Decision-makers also need to consider the costs and benefits associated with pursuing different strategies (see also Chapter 6), ranging from possible co-benefits to opportunity costs (Brown et al., 2011; Carter and Mäkinen, 2011; Smit et al., 1999; Wilbanks and Sathaye, 2007). It is important for decision-makers to consider that climate change will not be the only type of change confronting organizations in the future. Many organizations, industries and communities will be exposed to a variety of political, economic, socio-demographic and other types of environmental changes. These changes may pose, at times, additional difficulties in pursuing climate-resilient pathways. For instance, economic downturns and financial crises often draw attention away from longer-term issues associated with climate change or other types of global environmental changes (see Chapter 1).

Incremental and transformational strategies

Based on the findings of the IPCC Assessment Reports, it can be expected that different types of strategic responses in terms of adaptation, mitigation and resilience will be required depending on the magnitude and rate of climate change. The magnitude and rate of climate change is in turn dependent on the timing and levels of global GHG emissions and mitigation efforts, particularly in the second half of the twenty-first century and beyond (Peters et al., 2013). If the magnitude and rate of climate change can be kept to moderate levels due to stringent mitigation efforts, incremental adaptation may be a sufficient response to deal with the resulting impacts of climate change across many sectors and locations. However, in the absence of stringent mitigation efforts, and in industries and locations where vulnerabilities are already high, transformational approaches are likely needed to respond to the impacts from climate change (Denton et al., 2014; Kates et al., 2012). The implementation of such approaches ultimately requires overcoming path dependencies (that is, limiting circumstances that have been established through past decisions) and constraining institutional arrangements (that is, regulatory structures or prevailing social norms) (Adger et al., 2005; David, 1994).

Incremental and transformational strategies to prepare for the impacts of climate change are not mutually exclusive, but strategically very different. Incremental strategies consist of an organization's gradual reactions to immediate or anticipated future changes (Courtney et al.,

1997; Linnenluecke and Griffiths, 2010). These strategies are often based on, or an extension of, existing actions, behaviors and management approaches to reduce losses or exploit opportunities associated with climate change (Kates et al., 2012). For instance, organizations may choose to respond to climate change and policy developments by capitalizing on efficiency gains and include more energy-efficient practices to mitigate climate change impacts. Other organizations may choose to incorporate sea level rise projections into the location and design of new building developments. Incremental strategies are often referred to as 'business-as-usual' approaches, as they do not challenge or disrupt existing actions, behaviors and management approaches (Denton et al., 2014).

Transformational strategies, on the other hand, include actions, behaviors and management approaches that go beyond 'business-as-usual' approaches. Transformational strategies are not limited to the organizational level, but involve political, regulatory, economic, financial, social or technological systems. Transformational strategies can be defined as those strategies 'that are adopted at a much larger scale, are truly new to a particular region or resource system, and that transform places and shift locations' (Kates et al., 2012: 7156). Transformations in political, regulatory or economic systems can contribute to enhanced climate responses for adaptation, mitigation and resilience (Denton et al., 2014). Climate-resilient pathways are likely to involve significant transformations to be able to address adaptation, mitigation and resilience challenges. Examples for transformational strategies cited in the literature (see Enserink, 2004; Kates et al., 2012; Vis, Klijn et al., 2003) are the coastal defense program ('Weak Links') and the riverine flood abatement and water supply program ('Room for the River') in the Netherlands. Both programs focus on building adaptation and resilience to sea level rise, storm surges and changed patterns of precipitation. The riverine program represents a major variation to traditional Dutch flood control through levees, and instead focuses on enlarging flood plains, addressing spatial planning and improving ecological concerns. Together with low-carbon economic developments, such programs can represent a first step towards creating climate-resilient pathways.

Transformational change, especially across scales, as in the Dutch example, is often difficult to achieve, and there are many examples of failed attempts of transformations, restructuring, or reengineering efforts (Kotter, 1995). Transformational strategies are easily hampered by institutional, social or cultural barriers, including uncertainties about costs and future benefits (Kates et al., 2012). Transformational change can threaten existing worldviews of how things 'should be done' as well as vested interests by certain actors, such as access rights, or privileges (Kates et al., 2012; Kemp and Loorbach, 2006). Transformational change can also prioritize the interests of some over those of others (Denton et al., 2014).

Nonetheless, existing examples show that it is possible to support deliberate transformations to reduce the impacts of climate change and vulnerability (see Chapter 5), and to contribute to sustainable outcomes. Kates et al. (2012) argue that supportive social contexts are needed in combination with incentives, the availability of acceptable options, and resources for actions. In particular, the authors argue that an incorporation of transformation early on in planning and risk management, along with discussing 'what if' scenarios (see also Chapter 5), facilitates transformative change. The consideration of a range of possible innovations and options can thereby help to expand available choices. In addition, Denton et al. (2014) suggest that transformations can also be facilitated through governance mechanisms that promote innovations, alternative options, or new behaviors. Finally, transformations can also be facilitated when focal events (such as the impacts of climate or weather extremes) draw attention to the need for transformative change. Such events can mobilize groups or networks to advocate in favor of transformational change (Denton et al., 2014; Hernes, 2012).

Avoiding trade-offs and creating synergies
While adaptation, mitigation and resilience have the potential to contribute to successfully addressing challenges associated with future sustainable development, trade-offs may (Tol, 2005) – especially between pursuing economic goals and environmental goals. For instance, the maximization of economic returns through over-farming and over-fishing can easily lead to a loss of biodiversity and the over-exploitation of ecosystem services in the long run. These trends can be exacerbated by the impacts of climate change, with negative impacts for associated industries and communities.

Trade-offs between adaptation, mitigation and resilience are also possible. Adaptation to hotter days and heat waves in the building sector may lead to increased GHG emissions and expenditure if adaptation is achieved via increases in air-conditioning that uses non-renewable energy sources. In other cases, mitigation attempts may impede resilience. Some attempts at GHG mitigation, such as the subsidization of the US ethanol industry, have the potential to compromise long-term resilience through both undesirable ecological effects (such as the loss of crop diversity, along with soil erosion and groundwater depletion) and social effects (such as the reduction of flexibility for alternative fuel development and the potential creation of food insecurity) (Adger et al., 2011). Likewise, attempts at building adaptation and resilience can have negative side-effects. For instance, in central Vietnam, the construction of dams to prevent flooding and generate power threatens the livelihood of poor communities due to issues around relocating communities and inundating forestland to build dams (Beckman, 2011).

Particular challenges for building synergies between mitigation, adaptation and resilience arise due to the requirement of 'additionality' in climate change policy, that is, the requirement that the GHG emissions after the implementation of a mitigation project must be lower than those that would have occurred without implementation of the mitigation project (see Chapter 3). This requirement does often not fully consider trade-offs between adaptation to climate change, mitigating GHG emissions and building resilience. However, adaptation efforts can take into account the critical role of *co-benefits* and include opportunities to support mitigation and resilience goals, while at the same time reducing vulnerabilities to climate change impacts (Denton et al., 2014). Positive synergies thus need to be carefully considered, and may decrease once the limits to mitigating climate change are exceeded.

Ultimately, it might not be possible to alleviate all trade-offs between different strategies. Consequently, it will be important to understand how positive feedback loops between adaptation, mitigation, resilience and sustainable development can be enhanced with a view towards possibly achieving win-win situations while minimizing potential trade-offs (Denton et al., 2014). Previous attempts to achieve win-win situations have included the Clean Development Mechanism (CDM) and Joint Implementation (JI) (see Chapter 3). The CDM and JI have the aim to help mitigate carbon emissions while building capacities in local communities within both developed and developing countries. However, whether or not these approaches have led to win-win situations has been subject to debate. In the case of the CDM, findings have shown that generating sustainable development outcomes has been difficult. Sutter and Parreño (2007) analyzed 16 officially-registered CDM projects to determine whether they fulfill their objectives (as required by the Kyoto Protocol) in regards to GHG emission reductions and a contribution to sustainable development in the host country. They concluded that while 72 per cent of expected certified emission reductions (CERs) of the analyzed CDM projects are likely to lead to measurable emission reductions, only 1 per cent of the expected CERs are likely to contribute to a *significant* level of sustainable development within the host country.

An easier integration of different strategies might primarily occur in sectors such as urban and infrastructure planning, as well as in agriculture and forestry (Smith et al., 2007). For instance, infrastructure planning can include considerations in regards to selecting less vulnerable sites, climate-proofing infrastructure development, integrating energy conservation, paying attention to low carbon and adapted building characteristics (for example, building codes), alongside low-carbon infrastructure planning. The agriculture and forestry sectors, on the other hand, have opportunities to contribute to ecosystem resilience, prevent erosion, protect biodiversity, contribute to lower water evaporation and heat stress, and sequester carbon – however, some trade-offs can exist between plantations best-suited for carbon sequestration and those best-suited to achieve other types of outcomes (Smith et al., 2007). Therefore, there appears to be a growing need to consider multiple criteria for evaluating the success of any single strategy or action to respond to the impacts of climate change.

Issues of scale

Last, a key factor in integrating climate change adaptation, mitigation and resilience strategies is to consider differences in temporal and spatial scales (see Chapter 1). Costs and benefits affect different economic sectors in different ways, so that costs and benefits of adaptation, mitigation and resilience are distributed differently (Smith et al., 2007). The costs of mitigation efforts are largely localized, while the benefits of mitigation are global. On the other hand, the benefits of implementing adaptation and resilience strategies will be largely localized, however, larger transformational efforts often depend on support from a regional, national or even global scale. Looking at a temporal scale, mitigation decisions would have to be taken immediately and collectively by major emitters in order to significantly reduce the impacts of climate change. Direct responsibilities to curb the main drivers of global climate change are dispersed across countries. In contrast, adaptation often falls to practitioners where local responsibility is clearer, yet outcomes are more difficult to assess due to cost-benefit asymmetries – that is, near-term investments are not leading to near-term outcomes (Denton et al., 2014, see also Chapter 1).

Actors affecting and affected by adaptation, mitigation and resilience strategies are also different and have different levels of decision-making power and concentration. Mitigation decisions often affect powerful industrial stakeholders (such as those across the energy, transportation and extractive industries) with significant potential to influence policy decisions, while adaptation often involves more dispersed stakeholders at the local level across sectors such as urban planning, water, agriculture, health or coastal zone management (Smith et al., 2007; Wilbanks et al., 2007a). Especially in developing countries, local stakeholders are often limited in their capabilities to implement adaptation actions and influence policy decisions. Ultimately, issues surrounding scale are not easily addressed by single actors; however, they are likely to become an area of increasing importance. Currently policy discussions are often not considering these intricate scale issues (see also Chapters 3 and 4). However, future policy developments can be directed towards providing incentives for overcoming these issues.

Climate-Resilient Pathways

The challenge of embarking on climate-resilient pathways is one that will be difficult to meet. Nonetheless, actions, behaviors and management approaches can be pursued now that will contribute significantly to embarking on climate-resilient pathways. Swart and Raes (2007) suggest a number of factors that should be taken into consideration when looking into combining adaptation/resilience and mitigation. The authors see the following points as priorities:

> (1) avoiding trade-offs – when designing policies for mitigation or adaptation, (2) identifying synergies, (3) enhancing response capacity, (4) developing institutional links between adaptation and mitigation – e.g. in national institutions and in international negotiations, and (5) mainstreaming adaptation and mitigation considerations into broader sustainable development policies. (Swart and Raes, 2007: 288)

The challenge (not just for businesses and other organizations, but also for policy decisions and society) in pursuing climate-resilient pathways is, therefore, the identification and implementation of the right combination of technological and non-technological options that reduce net carbon emissions while supporting other corporate objectives, such as adaptation or the achievement of corporate sustainability outcomes (Denton et al., 2014).

In many sectors, especially those that are highly infrastructure dependent, such strategies are easiest to be implemented at the inception of a project. Evidence from the EU shows how such outcomes are achieved for the energy and transport infrastructure sectors as well as for the building sector. These sectors were also given priority in the 2009 White Paper on Climate Change Adaptation. Equipping energy, transportation and building infrastructure to cope with changing temperatures and to be resilient to weather extremes is crucial to guarantee a reliable infrastructure for the future. Mitigation measures (for example, the uptake of renewable energy and energy efficiency in buildings) can make important contributions to deal with the transition of energy systems and managing energy demand. While infrastructure projects that integrate adaptation, mitigation and resilience often have higher up-front capital costs, even if their climate resilience makes them more profitable over the lifetime of the infrastructure, they also open up new economic opportunities in training and skills development. In Europe, public-private partnership structures (PPP) are emerging as popular models for funding infrastructure investment (United Nations, 2000).

Enhancing the range of choices through innovation
The development of climate-resilient pathways should also consider possibilities to develop new options through innovation. For example, if a certain sector faces issues around water scarcity, the development of a climate-resilient pathway should take into consideration how innovation might lead to changes in existing water usage behaviors and allocation practices, improvements in groundwater reservoirs, as well as improved technologies and policies for efficiency improvements (Millennium Ecosystem Assessment, 2005a). Similarly, innovations in sectors such as agriculture can contribute to both mitigation and adaptation outcomes. For instance, practices such as conservation tillage can help to reduce carbon dioxide emissions, while contributing to increases in soil organic carbon and improvements in soil structure (IISD, 2006). The development and proliferation of new technologies also allows developing countries to develop social and institutional infrastructure to support a trajectory towards more climate-resilient pathways. For instance, access to mobile technologies in developing countries allows farmers to access vital information regarding disease surveillance, agricultural inputs and market prices for crops (Denton et al., 2014; Lenton and Schellnhuber, 2007).

New toolkits for assessing strategic planning
The success or failure of any strategic plan is ultimately highly dependent on whether a decision-maker truly understands the interaction and complexity of the system he or she is trying to influence (Radzicki and Taylor, 1997). When evaluating the opportunities and trade-offs associated with adaptation, mitigation and resilience, there is a need to combine different valuation metrics and information requirements, including those discussed in previous chapters. Key challenges are: weighting different priority areas, identifying adaptation, mitigation and resilience objectives,

identifying options and establishing decision-making criteria, and appraising options. Particular attention should be paid to identifying major trade-offs and potential synergies, as well as the implications which may vary according to climate impacts at specific locations, corporate vulnerabilities, as well as different temporal and spatial scales (Denton et al., 2014). Considering the complexity and size the issues that public and private sector decision-makers must manage, it is not surprising that the 'intuitive' or 'common sense' approach to strategic planning often falls short or is counter-productive (Radzicki and Taylor, 1997).

To assist decision-makers with the development of more sophisticated models and tools for strategic planning that consider greater levels of complexity, the literature has started to offer tools that can help understand, evaluate, visualize and map the behavior of more complex decision-making backgrounds. Tools include, for example, the application of systems thinking and organizational planning to improve organizational learning outcomes (Trenberth et al., 2007; van Vuuren et al., 2011). Other attempts include the development of agent-based models, which involves building computational models reflecting so-called 'agents' (reflecting actors in the real world) and their interactions, for example, to understand how different actors within a system would react to policy changes or pricing signals (IPCC, 2013; IPPC, 2007), as well as the use of Real Options Valuation (or Real Options Analysis) to evaluate organizational decisions under uncertainty (Bodansky, 2001; Grant, 2000; Hegerl et al., 2007).

The Role of Institutional Support

While the focus of this book has mainly been on the organizational level, the development of climate resilient pathways is greatly influenced by the surrounding institutional environment. Organizations will benefit from an institutional environment that is effective and creates enabling conditions to cope with a wide range of major future challenges (Gupta et al., 2010). As outlined in Chapter 1, climate change is only one of these major future challenges, alongside ocean acidification, stratospheric ozone concentration, the biogeochemical nitrogen (N) and phosphorus (P) cycles, global freshwater use, land-use system change, and the rate at which biological diversity is lost (Rockström et al., 2009a,b). In addition, and as outlined by Denton et al. (2014), better institutions are needed to support international efforts at adaptation and mitigation, and especially for managing the significant flow of funds and other resources directed at different actions to address climate change. There is currently a significant level of complexity surrounding resource flows and distributional consideration for programs aimed at adaptation, mitigation, and resilience (see Chapters 3 and 4). Effective institutional arrangements are necessary for common pool resources such as freshwater or fisheries, which have experienced common problems with institutional and governance arrangements. Management of these challenges can only be accomplished where there are strong institutions on multiple levels acting as stewards of a safe living space for humanity. The way forward will, therefore, need institutional arrangements that foster innovation, monitoring, and the evaluation of strategies for reducing the rate of climate change, managing climate impacts and reducing risks.

References

Adger, N. W., Arnell, N. W., and Tompkins, E. L. 2005. 'Successful adaptation to climate change across scales'. *Global Environmental Change*, **15**(2): 77–86.

Adger, W. N. 2006. 'Vulnerability'. *Global Environmental Change*, **16**(3): 268–81.

Adger, W. N., Agrawala, S., Mirza, M. M. Q., Conde, C., O'Brien, K., Pulhin, J., and Co-Authors. 2007. 'Assessment of adaptation practices, options, constraints and capacity'. In M. L. Parry, O. F. Canziani, J. P. Palutikof, P. J. van der Linden, and C. E. Hanson (eds), *Climate Change 2007: Impacts, Adaptation and Vulnerability. Contribution of Working Group II to the Fourth Assessment Report of the Intergovernmental Panel on Climate Change*. Cambridge: Cambridge University Press.

Adger, W. N., Brown, K., Nelson, D. R., Berkes, F., Eakin, H., Folke, C., and Co-Authors. 2011. 'Resilience implications of policy responses to climate change'. *Wiley Interdisciplinary Reviews: Climate Change*, **2**(5): 757–66.

Adger, W. N., Huq, S., Brown, K., Conway, D., and Hulme, M. 2003. 'Adaptation to climate change in the developing world'. *Progress in Development Studies*, **3**(3): 179.

Agrawala, S., and Fankhauser, S. 2008. *Economics Aspects of Adaptation to Climate Change. Costs, Benefits and Policy Instrument*. Paris: OECD.

Aldy, J., and Stavins, R. 2012. *Climate Negotiations Open a Window: Key Implications of the Durban Platform for Enhanced Action*. Cambridge, MA: Harvard Kennedy School.

Allen, M. R. 2010. 'Attributing extreme weather events: implications for liability'. In Munich Re (ed.), *Liability for Climate Change? Experts' Views on Potential Emerging Risk*. Munich, Germany: Münchener Rückversicherungs-Gesellschaft.

Allenby, B., and Fink, J. 2005. 'Toward inherently secure and resilient societies'. *Science*, **309**(5737): 1034–36.

Alley, R. B., Marotzke, J., Nordhaus, W. D., Overpeck, J. T., Peteet, D. M., Pielke Jr, R. A., and Co-Authors. 2003. 'Abrupt climate change'. *Science*, **299**(5615): 2005–10.

Anglian Water 2011. *Climate Change Adaptation Report.* Accessed 24 September 2014, available at: http://www.anglianwater.co.uk/_assets/media/CC_Adaptation_Report_Final_compressed_corrected.pdf.

Anglian Water 2014. *Adapting to Climate Change.* Accessed 24 September 2014, available at: http://www.anglianwater.co.uk/environment/climate-change/adaption.aspx.

Arauzo-Carod, J. M., Liviano-Solis, D., and Manjón-Antolín, M. 2010. 'Empirical studies in industrial location: an assessment of their methods and results'. *Journal of Regional Science*, **50**: 685–711.

Arnell, N. 2013. *The Scale of the Challenge: Uncertainty, the Effects of Mitigation Policy and the Implications for Adaptation.* Paper presented at the European Climate Change Adaptation Conference, Hamburg. Accessed 24 September 2014, available at: http://www.walker-institute.ac.uk/events/Arnell-ECCA-presentation.pdf.

Attavanich, W., McCarl, B. A., Ahmedov, Z., Fuller, S. W., and Vedenov, D. V. 2013. 'Effects of climate change on US grain transport'. *Nature Climate Change*, **3**: 638–43.

Australian Bureau of Statistics 2003. *Queensland in Review.* Accessed 24 September 2014, available at: http://www.abs.gov.au/Ausstats/abs@.nsf/Lookup/CF3424B58ECB69C8CA256CC500211FCA.

Barrett, S. 2008. 'Climate treaties and the imperative of enforcement'. *Oxford Review of Economic Policy*, **24**(2): 239–58.

Bateman, I. J., Mace, G. M., Fezzi, C., Atkinson, G., and Turner, K. 2011. 'Economic analysis for ecosystem service assessments'. *Environmental and Resource Economics*, **48**(2): 177–218.

Bates, J. 2013. 'Big data lets global corps bet on the threat of climate change'. Accessed 24 September 2014, available at: http://theconversation.com/big-data-lets-global-corps-bet-on-the-threat-of-climate-change-19501. *The Conversation.*

Beckman, M. 2011. 'Converging and conflicting interests in adaptation to environmental change in central Vietnam'. *Climate and Development*, **3**(1): 32–41.

Bellamy, R., and Aron, H. 2010. *Climate Adaptation Tool.* Accessed 24 September 2014, available at: http://www.norfolk.gov.uk/view/NCC095103.

Bishop, R. C. 1978. 'Endangered species and uncertainty: the economics of a safe minimum standard'. *American Journal of Agricultural Economics*, **60**(1): 10–18.

Bodansky, D. 2001. 'The history of the global climate change regime'. In U. Luterbacher, and D. Sprinz (eds), *International Relations and Global Climate Change*. Cambridge, MA: MIT Press.

Boin, A., and McConnell, A. 2007. 'Preparing for critical infrastructure breakdowns: the limits of crisis management and the need for resilience'. *Journal of Contingencies and Crisis Management*, **15**(1): 50–9.

Bradshaw, B., Dolan, H., and Smit, B. 2004. 'Farm-level adaptation to climatic variability and change: Crop diversification in the Canadian prairies'. *Climatic Change*, **67**(1): 119–41.

Brooks, N., and Adger, W. N. 2003. *Country Level Risk Measures of Climate-Related Natural Disasters and Implications for Adaptation to Climate Change*. Manchester: Tyndall Centre for Climate Change Research, Manchester University, UK.

Brouwer, A. E. 2010. 'The old and the stubborn? Firm characteristics and relocation in the Netherlands'. *European Spatial Research and Policy*, **17**: 41–60.

Brouwer, A. E., Mariotti, I., and van Ommeren, J. N. 2004. 'The firm relocation decision: an empirical investigation'. *The Annals of Regional Science,* **38**: 335–47.

Brown, A., Gawith, M., Lonsdale, K., and Pringle, P. 2011. *Managing Adaptation: Linking Theory and Practice.* Oxford: UK Climate Impacts Programme.

Brown, M., and Funk, C. 2008. 'Food security under climate change'. *Science*, **319**: 580–81.

Brunner, S. F., C. , Luderer, G., and Edenhofer, O. 2011. *Emissions Trading Systems: An Overview.* Potsdam Institute for Climate Impact Research Discussion paper. Postdam, Germany.

Burby, R., Cigler, B., French, S., Kaiser, E., Kartez, J., Roenigk, D., and Co-Authors. 1991. *Sharing Environmental Risks: How to Control Governments' Losses in Natural Disasters.* Boulder, CO: Westview Press.

Burton, I., Huq, S., Lim, B., Pilifosova, O., and Schipper, E. 2002. 'From impacts assessment to adaptation priorities: the shaping of adaptation policy'. *Climate Policy*, **2**: 145–59.

Camerer, C., and Kunreuther, H. 1989. 'Decision processes for low probability events: policy implications'. *Journal of Policy Analysis and Management*, **8**(4): 565–92.

Cantley-Smith, R. 2010. 'Climate change and the Copenhagen legacy: where to from here'. *Monash University Law Review*, **36**: 278.

Carbon Disclosure Project. 2012. *Insights into Climate Change Adaptation by UK Companies.* Accessed 24 September 2014, available at: http://archive.defra.gov.uk/environment/climate/documents/cdp-adaptation-report.pdf. London: Carbon Disclosure Project.

Carson, R. 1962. *Silent Spring.* Boston, MA: Houghton Mifflin.

Carter, T. R., and Mäkinen, K. 2011. *Approaches to Climate Change Impact, Adaptation and Vulnerability Assessment: Towards a Classification Framework to Serve Decision-Making.* MEDIATION Technical Report No. 2.1. Helsinki: Finnish Environment Institute (SYKE).

Carter, T. R., Jones, R. N., Lu, X., Bhadwal, S., Conde, C., Mearns, L. O., and Co-Authors. 2007. 'New assessment methods and the characterization of future conditions'. In M. L. Parry, O. F. Canziani, J. P. Palutikof, P. J. van der Linden, and C. E. Hanson (eds), *Climate Change 2007: Impacts, Adaptation and Vulnerability: Contribution of Working Group II to the Fourth Assessment Report of the Intergovernmental Panel on Climate Change*, Cambridge: Cambridge University Press.

Cash, D. W., Adger, W. N., Berkes, F., Garden, P., Lebel, L., Olsson, P., and Co-Authors. 2006. 'Scale and cross-scale dynamics: governance and information in a multilevel world'. *Ecology and Society*, **11**(2): 8.

Center for Climate and Energy Solutions 2011. *Outcomes of the UN Climate Change Conference in Durban, South Africa.* Accessed 24 September 2014, available at: http://www.c2es.org/docUploads/COP17_Summary.pdf.

Center for Climate and Energy Solutions 2012. *Outcomes of the UN Climate Change Conference in Doha, Qatar.* Accessed 24 September 2014, available at: http://www.c2es.org/docUploads/c2es-cop-18-summary.pdf.

Center for Climate and Energy Solutions 2013. *Outcomes of the UN Climate Change Conference in Warsaw.* Accessed 24 September 2014, available at: http://www.c2es.org/international/negotiations/cop-19/summary.

Center for Climate and Energy Solutions 2014. *State and Local Climate Adaptation.* Accessed 24 September 2014, available at: http://www.c2es.org/us-states-regions/policy-maps/adaptation.

Chambwera, M., Heal, G., Dubeux, C., Hallegatte, S., Leclerc, L., Markandya, A., and Co-Authors. 2014. Economics of Adaptation. In IPCC (ed.), *Climate Change 2014: Impacts, Adaptation, and Vulnerability: Contribution of Working Group II to the Fifth Assessment Report of the Intergovernmental Panel on Climate Change.* Accessed 24 September 2014, available at: http://ipcc-wg2.gov/AR5/report/.

Coase, R. 1960. 'The problem of social cost'. *Journal of Law and Economics*, **3**: 1–44.

Coffee, J. 2014. 'Climate change a growing concern for companies expanding their footprint'. Accessed 24 September 2014, available at http://www.theguardian.com/sustainable-business/hubs-water-climate-change-siting-drought-flood-business. *The Guardian.*

Commission of the European Communities 2009. *White Paper – Adapting to Climate Change: Towards a European Framework for Action.* Accessed 24 September 2014, available at: http://eur-lex.europa.eu/ LexUriServ/LexUriServ.do?uri=COM:2009:0147:FIN:EN:PDF. Brussels.

Commonwealth Scientific and Industrial Research Organisation (CSIRO) 2009. *The Science of Tackling Climate Change.* Accessed 30 September 2014, available at: http://www.csiro.au/~/media/CSIROau/Divisions/ CSIRO%20Ecosystem%20Sciences/ScienceTacklingClimateChange_ CSE_PDF%20Standard.pdf.

Confino, J. 2013. 'What's the purpose of sustainability reporting?' Accessed 24 September 2014, available at: http://www.guardian.co.uk/ sustainable-business/blog/what-is-purpose-of-sustainability-reporting. *The Guardian.*

Cook, J., Nuccitelli, D., Green, S. A., Richardson, M., Winkler, B., Painting, R., and Co-Authors. 2013. 'Quantifying the consensus on anthropogenic global warming in the scientific literature'. *Environmental Research Letters*, **8**(2): 024024.

Courtney, H. 2003. 'Decision-driven scenarios for assessing four levels of uncertainty'. *Strategy and Leadership*, **31**(1): 14–22.

Courtney, H., Kirkland, J., and Viguerie, P. 1997. 'Strategy under uncertainty'. *Harvard Business Review*, **75**(6): 67–79.

Dales, J. 1968. *Pollution, Property and Prices.* Toronto, ON: University Press.

Daly, H. E. (ed.) 1973. *Toward a Steady-State Economy.* San Francisco, CA: W. H. Freeman.

Daly, H. E. 1974. 'The world dynamics of economic growth: the economics of the steady state'. *American Economic Review*, **64**(2): 15–21.

Daly, H. E. 1993. 'Sustainable growth: an impossibility theorem'. In H. Daly, and K. Townsend (eds), *Valuing the Earth: Economics, Ecology Ethics.* Cambridge, MA: MIT Press.

Dargusch, P., and Griffiths, A. 2008. 'Introduction to special issue: a typology of environmental markets'. *Australasian Journal of Environmental Management*, **15**(2): 70–75.

David, P. A. 1994. 'Why are institutions the "carriers of history"? Path dependence and the evolution of conventions, organizations and institutions'. *Structural Change and Economic Dynamics*, **5**(2): 205–20.

De Boer, J., Wardekker, J. A., and Van der Sluijs, J. P. 2010. 'Frame-based guide to situated decision-making on climate change'. *Global Environmental Change*, **20**(3): 502–10.

De Loë, R., Kreutzwiser, R., and Moraru, L. 2001. 'Adaptation options for the near term: climate change and the Canadian water sector'. *Global Environmental Change*, **11**(3): 231–45.

DEFRA (Department for Environment Food and Rural Affairs) 2012. *UK Climate Change Risk Assessment: Government Report.* London: The Stationery Office (TSO) Ltd.

DEFRA (Department for Environment Food and Rural Affairs) 2013. *The National Adaptation Programme: Making the Country Resilient to a Changing Climate.* London: The Stationery Office (TSO) Ltd.

Denton, F., Wilbanks, T., Abeysinghe, A. C., Burton, I., Gao, Q., Lemos, M. C., and Co-Authors. 2014. 'Climate-resilient pathways: adaptation, mitigation, and sustainable development'. In IPCC (ed.), *Climate Change 2014: Impacts, Adaptation, and Vulnerability: Contribution of Working Group II to the Fifth Assessment Report of the Intergovernmental Panel on Climate Change.* Accessed 24 September 2014, available at: http://ipcc-wg2.gov/AR5/report/.

Department of Climate Change 2009. *Climate Change Risks to Australia's Coast: A First Pass National Assessment.* Accessed 24 September 2014, available at: http://www.climatechange.gov.au/sites/climate change/files/documents/03_2013/cc-risks-full-report.pdf.

Department of Climate Change and Energy Efficiency 2007. *National Climate Change Adaptation Framework.* Accessed 24 September 2014, available at: http://www.climatechange.gov.au/sites/climatechange/files/documents/03_2013/nccaf.pdf.

Department of Climate Change and Energy Efficiency 2010. *Developing a National Coastal Adaptation Agenda: A Report on the National Climate Change Forum.* Accessed 24 September 2014, available at: http://www.climatechange.gov.au/sites/climatechange/files/documents/03_2013/developing-national-coastal-adaptation-agenda.pdf.

Department of Climate Change and Energy Efficiency 2012. *Economic Framework for Analysis of Climate Change Adaptation Options Framework Specification.* Accessed 24 September 2014, available at: http://www.climatechange.gov.au/sites/climatechange/files/documents/08_2013/economic-framework-adaptation-options.pdf.

Department of Primary Industries and Fisheries 2006. *Cyclone Larry: Briefing Note.* Brisbane: Department of Primary Industries and Fisheries.

Department of the Environment 2014. 'Community discussion'. Accessed 24 September 2014, available at: http://www.climatechange.gov.au/community-discussion.

Department of the Environment and Heritage, Australian Greenhouse Office 2013. *Climate Change Impacts and Risk Management: A Guide for Business and Government.* Accessed 24 September 2014, available at http://www.climatechange.gov.au/sites/climatechange/files/documents/03_2013/risk-management.pdf.

Dessai, S., and Hulme, D. 2004. 'Does climate adaptation policy need probabilities?'. *Climate Policy* **4**: 107–28.

Dow, K., Berkhout, F., Preston, B. L., Klein, R. J., Midgley, G., and Shaw, M. R. 2013. 'Limits to adaptation'. *Nature Climate Change*, **3**(4): 305–7.

Dunphy, D. C., and Griffiths, A. 1998. *The Sustainable Corporation: Organisational Renewal in Australia*. St Leonards, Australia: Allen and Unwin.

Dunphy, D. C., Griffiths, A., and Benn, S. 2003. *Organizational Change for Corporate Sustainability: A Guide for Leaders and Change Agents of the Future*. London: Routledge.

Economics of Climate Adaptation Working Group 2009. *Shaping Climate-Resilient Development: A Framework for Decision-Making*. Accessed 24 September 2014, available at ccsl.iccip.net/climate_resilient.pdf.

EG Science. 2008. *The 2°C Target Information Reference Document: Background on Impacts, Emission Pathways, Mitigation Options and Costs*. Prepared and adopted by EU Climate Change Expert Group *EG Science*.

Egri, C. P., and Pinfield, L. (eds). 1996. *Organizations and the Bio-sphere: Ecologies and Environments*. Newbury Park, CA: Sage.

Ehrenfeld, J. R. 2005. 'Eco-efficiency'. *Journal of Industrial Ecology*, **9**(4): 6–8.

Ehrlich, P. R. 1968. *The Population Bomb*. New York: Ballentine Books.

Elkington, J. 1997. *Cannibals with Forks: The Triple Bottom Line of 21st Century Business*. Oxford: Capstone.

Enserink, B. 2004. 'Thinking the unthinkable – the end of the Dutch river dike system? Exploring a new safety concept for the river management'. *Journal of Risk Research*, **7**(7–8): 745–57.

Entergy 2005. *Annual Report*. Accessed 24 September 2014, available at: http://www.entergy.com/content/investor_relations/pdfs/2005ARFINAL.pdf.

European Commission 2009. *White Paper. Adapting to Climate Change: Towards a European framework for action*. Brussels: Commission of the European Communities.

European Commission 2012. *The State of the European Carbon Market in 2012: Report from the Comission to the European Parliament and the Council*. Brussels: Commission of the European Communities.

European Commission 2013. *The EU Strategy on Adaptation to Climate Change*. Accessed 24 September 2014, available at: http://ec.europa.eu/clima/publications/docs/eu_strategy_en.pdf.

European Commission 2014. *Structural Reform of the European Carbon Market*. Accessed 24 September 2014, available at: http://ec.europa.eu/clima/policies/ets/reform/index_en.htm.

European Environment Agency 2010. *The European Environment – State and Outlook 2010*. Accessed 24 September 2014, available at: www.eea.europa.eu/soer.

ExxonMobil 2014. *Report: Energy and Carbon – Managing the Risks*. Accessed 24 September 2014, available at: http://corporate.exxon mobil.com/en/environment/climate-change/managing-climate-change-risks/carbon-asset-risk.

Fankhauser, S., Smith, J., and Tol, R. 1999. 'Weathering climate change: some simple rules to guide adaptation decisions'. *Ecological Economics*, **30**: 67–78.

Feenstra, J., Burton, I., Smith, J., and Tol, R. (eds). 1998. *Handbook on Methods for Climate Change Impact Assessment and Adaptation Strategies*. Amsterdam: UNEP/Vrije Universiteit.

Felgenhauer, T., and Webster, M. 2013. 'Multiple adaptation types with mitigation: a framework for policy analysis'. *Global Environmental Change*, **23**(6): 1556–65.

Fischer, K., and Schot, J. 1993. *Environmental Strategies for Industry: International Perspectives on Research Needs and Policy Implications*. Washington, DC: Island Press.

Fleming, A., Hobday, A., Farmery, A., van Putten, E., Pecl, G., Green, B., and Lim-Camacho, L. 2014. 'Climate change risks and adaptation options across Australian seafood supply chains – A preliminary assessment'. *Climate Risk Management*, 1: 39–50.

Foley, J. A., Ramankutty, N., Brauman, K. A., Cassidy, E. S., Gerber, J. S., Johnston, M., and Co-Authors. 2011. 'Solutions for a cultivated planet'. *Nature*, **478**(7369): 337–42.

Folke, C., Hahn, T., Olsson, P., and Norberg, J. 2005. 'Adaptive governance of social-ecological systems'. *Annual Review of Environment and Resources*, 30: 441–73.

Foxon, T., and Andersen, M. M. 2009. *The Greening of Innovation Systems for Ecoinnovation: Towards an Evolutionary Climate Mitigation Policy*. Paper presented at the DRUID Summer Conference – Innovation, Strategy and Knowledge, Copenhagen Business School, Denmark.

Füssel, H., and Klein, R. 2006. 'Climate change vulnerability assessments: an evolution of conceptual thinking'. *Climatic Change*, 75: 301–29.

Gardner, J., Parsons, R., and Paxton, G. 2010. *Adaptation Benchmarking Survey: Initial Report*: Canberra: CSIRO Climate Adaptation National Research Flagship.

Garnaut, R. 2011. *The Garnaut Review 2011: Australia in the Global Response to Climate Change.* Cambridge: Cambridge University Press.

Geneva Association 2013. *Warming of the Oceans and Implications for the (Re)insurance Industry.* Geneva, Switzerland: Geneva Association.

Gibson, C. C., Ostrom, E., and Ahn, T.K. 2000. 'The concept of scale and the human dimensions of global change: a survey'. *Ecological Economics*, **32**(2): 217–39.

Gilding, P. 2011. *The Great Disruption: How the Climate Crisis will Change Everything (for the Better).* New York: Bloomsbury Publishing.

Giorgi, F. 2006. 'Climate change hot-spots'. *Geophysical Research Letters*, **33**(8): 1-4.

Gittell, J., Cameron, K., Lim, S., and Rivas, V. 2006. 'Relationships, layoffs, and organizational resilience airline industry responses to September 11'. *The Journal of Applied Behavioral Science*, **42**(3): 300–29.

Gladwin, T. N. 1993. 'The meaning of greening: a plea for organizational theory'. In K. Fischer, and J. Schot (eds), *Environmental Strategies for Industries: International Perspectives on Research Needs and Policy Implications.* Washington, DC: Island Press.

Gledhill, R., Hamza-Goodacre, D., and Ping Low, L. 2013. *Business-not-as-usual: Tackling the Impact of Climate Change on Supply Chain Risk.* Accessed 24 September 2014, available at http://www.pwc.com/ en_GX/gx/governance-risk-compliance-consulting-services/resilience/ publications/pdfs/issue3/business_not_as_usual.pdf.

Goulden, M., Conway, D., and Persechino, A. 2009. 'Adaptation to climate change in international river basins in Africa: a review'. *Hydrological Sciences*, **54**(5): 805–28.

Grant, L. 2000. *Too Many People: The Case for Reversing Growth.* Santa Ana, CA: Seven Locks Press.

Griffiths A., and Linnenluecke, M. K. 2008. 'Building corporate sustainability'. In Committee for Economic Development of Australia (CEDA) (eds), *Sustainable Queensland: Volume 3.* Melbourne: Committee for Economic Development of Australia.

Grubb, M., and Neuhoff, K. 2006. 'Allocation and competitiveness in the EU emissions trading scheme: policy overview'. *Climate Policy*, 6(1): 7–30.

Gunderson, L. H., and Holling, C. (eds). 2002. *Panarchy: Understanding Transformations in Human and Natural Systems.* Washington, DC: Island Press.

Gupta, J. 2010. 'A history of international climate change policy'. *Wiley Interdisciplinary Reviews: Climate Change*, 1(5): 636–53.

Hahn, T., Figge, F., Pinkse, J., and Preuss, L. 2010. 'Trade-offs in corporate sustainability: you can't have your cake and eat it'. *Business Strategy and the Environment*, **19**(4): 217–29.

Haines, A., Kovats, R. S., Campbell-Lendrum, D., and Corvalán, C. 2006. 'Climate change and human health: impacts, vulnerability and public health'. *Public Health*, **120**(7): 585–96.

Hall, A., and Jones, G. V. 2009. 'Effect of potential atmospheric warming on temperature-based indices describing Australian winegrape growing conditions'. *Australian Journal of Grape and Wine Research*, **15**(2): 97–119.

Hallegatte, S. 2008. 'An adaptive regional input-output model and its application to the assessment of the economic cost of Katrina'. *Risk Analysis*, **28**(3): 779–99.

Hallegatte, S. 2009. 'Strategies to adapt to an uncertain climate change'. *Global Environmental Change*, 19(2): 240–47.

Hallegatte, S., and Dumas, P. 2009. 'Can natural disasters have positive consequences? Investigating the role of embodied technical change. *Ecological Economics*', **68**(3): 777–86.

Hamin, E. M., and Gurran, N. 2009. 'Urban form and climate change: balancing adaptation and mitigation in the US and Australia'. *Habitat International*, **33**(3): 238–45.

Hannan, M. T., and Freeman, J. 1977. 'The population ecology of organizations'. *American Journal of Sociology*, 82: 929–64.

Hansen, J. E., and Sato, M. 2012. 'Paleoclimate implications for human-made climate change', *Climate Change*: 21–47: Springer.

Hardee, K., and Mutunga, C. 2010. 'Strengthening the link between climate change adaptation and national development plans: lessons from the case of population in national adaptation programmes of action (NAPAs)'. *Mitigation and Adaptation Strategies for Global Change*, **15**(2): 113–26.

Hardin, G. 1968. 'The tragedy of the commons'. *Science*, **162**(3859): 1243–48.

Hart, S. L. 1995. 'A natural-resource-based view of the firm'. *Academy of Management Review*: 986–1014.

Hart, S. L. 1997. 'Beyond greening: Strategies for a sustainable world'. *Harvard Business Review*: 66–76.

Hegerl, G., et al. 2007. 'Climate change 2007: the physical science basis'. In S. Solomon, et al. (eds), *Contribution of Working Group I to the Fourth Assessment Report of the Intergovernmental Panel on Climate Change*. Cambridge: Cambridge University Press.

Henriet, D., and Michel-Kerjan, E. 2008. *Looking at Optimal Risk-Sharing in a Kaleidoscope: The (Market Power, Information) Rotational Symmetry*. Philadelphia, PA: The Wharton School of the University of Pennsylvania.

Hernes, G. 2012. *Hot Topic-Cold Comfort: Climate Change and Attitude Change*. Oslo: Fafo Institute for Labour and Social Research.

Hertin, J., Berkhout, F., Gann, D. M., and Barlow, J. 2003. 'Climate change and the UK house building sector: perceptions, impacts and adaptive capacity'. *Building Research and Information*, **31**(3): 278–90.

Hess, M. 2004. '"Spatial" relationships? Towards a reconceptualization of embeddedness'. *Progress in Human Geography*, **28**: 165–86.

Hewitt, K., and Burton, I. 1971. *The Hazardousness of a Place: A Regional Ecology of Damaging Events*. Toronto, ON: University of Toronto Press.

Hintz, G., Normann, M., and Riese, J. 2011. *Get Adept at Adaptation*. Accessed 24 September 2014, available at: https://crawford.anu.edu.au/public_policy_community/workshops/climate_change_adaptation_in_australia/Get%20adept%20at%20adaptation%20McKinsey.pdf.

Hobday, A., Spillman, C., Hartog, J., and Hudson, D. 2012. *Adapting Primary Industries for Climate Change – Seasonal Forecasting as a Stepping Stone for Fisheries and Aquaculture*. Paper presented at the 2012 Australian Frontiers of Science Conference: Science for a Green Economy, Sydney, Australia.

Hoffman, A. J. 2001. *From Heresy to Dogma: An Institutional History of Corporate Environmentalism*. Stanford, CA: Stanford University Press.

Houghton, J. T. 2004. *Global Warming: The Complete Briefing* (3rd edn) Cambridge: Cambridge University.

House Standing Committee on Climate Change Water Environment and the Arts. 2009. *Managing our Coastal Zone in a Changing Climate: The Time to Act is Now*. Accessed 24 September 2014, available at: http://www.aph.gov.au/parliamentary_business/committees/house_of_representatives_committees?url=ccwea/coastalzone/report.htm.

Huppert, H., and Sparks, R. 2006. Extreme natural hazards: population growth, globalization and environmental change. *Philosophical Transactions of the Royal Society of London; Series A: Mathematical Physical and Engineering Sciences*, **364**(1845): 1875–88.

International Carbon Action Partnership (2014). *ETS Map*. Accessed 24 September 2014, available at: https://icapcarbonaction.com/ets-map.

International Institute for Sustainable Development (IISD) 2006. *The Sustainable Development Timeline*. Accessed 24 September 2014, available at: http://www.iisd.org/pdf/2006/sd_timeline_2006.pdf.

International Union of Railways 2012. *Adapting to Climate Changes.* Accessed 24 September 2014, available at: http://www.uic.org/spip.php?article1801.

IPCC 2001. *Climate Change 2001: Impacts, Adaptation and Vulnerability, Contribution of Working Group II to the Third Assessment Report of the Intergovernmental Panel on Climate Change.* Cambridge and New York: Cambridge University Press.

IPCC 2007a. *Climate Change 2007: Synthesis Report.* In R. Pachauri, and A. Reisinger (eds), *Contribution of Working Groups I, II and III to the Fourth Assessment Report of the Intergovernmental Panel on Climate Change.* Accessed 24 September 2014, available at: http://www.ipcc.ch/publications_and_data/ar4/syr/en/contents.html.

IPCC 2007b. 'Summary for policymakers.' In S. Solomon, D. Qin, M. Manning, Z. Chen, M. Marquis, K. B. Averyt, M. Tignor, and H. L. Miller (eds), *Climate Change 2007: The Physical Science Basis. Contribution of Working Group I to the Fourth Assessment Report of the Intergovernmental Panel on Climate Change.* Cambridge and New York: Cambridge University Press.

IPCC 2012. *Managing the Risks of Extreme Events and Disaster to Advance Climate Change Adaptation: Special Report of the Intergovernmental Panel on Climate Change.* Cambridge and New York: Cambridge University Press.

IPCC 2013. Summary for policymakers. In T. Stocker, D. Qin, G. Plattner, M. Tignor, S. Allen, J. Boschung, A. Nauels, Y. Xia, V. Bex, and P. Midgley (eds), *Climate Change 2013: The Physical Science Basis. Contribution of Working Group I to the Fifth Assessment Report of the Intergovernmental Panel on Climate Change.* Cambridge and New York: Cambridge University Press.

IPCC 2014a. Economics of Adaptation, *Climate Change 2014: Impacts, Adaptation, and Vulnerability. Contribution of Working Group II to the Fifth Assessment Report of the Intergovernmental Panel on Climate Change.* Accessed 24 September 2014, available at http://ipcc-wg2.gov/AR5/images/uploads/WGIIAR5-Chap17_FGDall.pdf.

IPCC 2014b. Introductory chapter, *Climate Change 2014: Mitigation of Climate Change. Contribution of Working Group III to the Fifth Assessment Report of the Intergovernmental Panel on Climate Change.* Accessed 24 September 2014, available at: http://report.mitigation2014.org/drafts/final-draft-postplenary/ipcc_wg3_ar5_final-draft_postplenary_chapter1.pdf.

IPCC 2014c. 'Summary for policymakers'. In C. Field, V. Barros, D. Dokken, K. Mach, M. Mastrandrea, T. C. Bilir, M. K. Ebi, Y. Estrada, R. Genova, B. Girma, E. Kissel, A. Levy, S. MacCracken, P. Mastrandrea, and L. White (eds), *Climate Change 2014: Impacts, Adaptation, and Vulnerability. Part A: Global and Sectoral Aspects. Contribution of Working Group II to the Fifth Assessment Report of the Inter-governmental Panel on Climate Change.* Cambridge and New York: Cambridge University Press.

Jha, A., Bloch, R., and Lamond, J. 2013. *Cities and Flooding: A Guide to Integrated Urban Flood Risk Management for the 21st Century.* Accessed 24 September 2014, available at: https://openknowledge. worldbank.org/handle/10986/2241.

Johnston, G., Burton, D., and Baker-Jones, M. 2013. *Climate Change Adaptation in the Boardroom.* Gold Coast, Australia: National Climate Change Adaptation Research Facility.

Jones, C. A., and Levy, D. L. 2007. 'North American business strategies towards climate change'. *European Management Journal,* **25**(6): 428–40.

Jones, R., and Boer, R. 2005. 'Assessing current climate risks'. In B. Lim, E. B. Spanger-Siegfried, I, E. Malone, and S. Huq (eds), *Adaptation Policy Frameworks for Climate Change: Developing Strategies, Policies and Measures.* Accessed 24 September 2014, available at: http://www.preventionweb.net/files/7995_APF.pdf#page=96, 91–118.

Julius, S., and Scheraga, J. 2000. 'The TEAM model for evaluating alternative adaptation strategies'. In Y. Haimes, and R. Steuer (eds), *Research and Practice in Multiple Criteria Decision Making.* New York: Springer-Verlag.

Kallio, T. J., and Nordberg, P. 2006. 'The evolution of organizations and natural environment discourse: some critical remarks'. *Organization & Environment,* **19**(4): 439–57.

Kates, R. W. 2000. 'Cautionary tales: adaptation and the global poor'. *Climatic Change,* **45**(1): 5–17.

Kates, R. W., Clark, W. C., Corell, R., Hall, J. M., Jaeger, C. C., Lowe, I., and Co-authors. 2001. 'Sustainability science'. *Science,* **292**: 641–42.

Kates, R. W., Travis, W. R., and Wilbanks, T. J. 2012. 'Transformational adaptation when incremental adaptations to climate change are insufficient'. *Proceedings of the National Academy of Sciences,* **109**(19): 7156–61.

Kelly, D., Kolstad, C., and Mitchell, G. 2005. 'Adjustment costs from environmental change. *Journal of Environmental Economics and Management'*, 50(3): 468–95.

Kemp, R., and Loorbach, D. 2006. 'Transition management: a reflexive governance approach'. In J.-P. Voss, D. Bauknecht, and R. Kemp (eds), *Reflexive Governance for Sustainable Development*: 103–30. Cheltenham, UK and Northampton, MA, USA: Edward Elgar.

Kidd, C. 1992. 'The evolution of sustainability'. *Journal of Agricultural and Environmental Ethics,* **5**(1): 1–26.

Klein, R. J. T., Nicholls, R. J., and Thomalla, F. 2003. 'Resilience to natural hazards: how useful is this concept?' *Global Environmental Change*, **5**(1–2): 35–45.

Knoben, J., and Oerlemans, L. 2008. 'Ties that spatially bind? A relational account of the causes of spatial firm mobility'. *Regional Studies*, **42**(3): 385–400.

Kotter, J. P. 1995. 'Leading change: why transformation efforts fail'. *Harvard Business Review*, **73**(2): 59–67.

Kriegler, E., Hall, J. W., Held, H., Dawson, R., and Schellnhuber, H. J. 2009. 'Imprecise probability assessment of tipping points in the climate system'. *Proceedings of the National Academy of Sciences*, **106**(13): 5041–46.

Laffont, J. 1995. 'Regulation, moral hazard and insurance of environmental risks'. *Journal of Public Economics*, **58**(3): 319–36.

LDC Expert Group 2012. *Best Practices and Lessons Learned in Adressing Adaptation in the Least Developed Countries*. Bonn, Germany: United Nations Climate Change Secretariat.

Lemmen, D., and Warren, F. (eds). 2004. *Climate Change Impacts and Adaptation: A Canadian Perspective*. Ottawa, Ontario: Natural Resources Canada.

Lenton, T. M., Held, H., Kriegler, E., Hall, J. W., Lucht, W., Rahmstorf, S., and Schellnhuber, H. J. 2008. 'Tipping elements in the Earth's climate system'. *Proceedings of the National Academy of Sciences*, 105(6): 1786–93.

Lenton, T., and Schellnhuber, H. 2007. 'Tipping the scales'. *Nature Reports Climate Change*, 1: 97–8.

Likens, G. E., and Bormann, F. H. 1974. 'Acid rain: a serious regional environmental problem'. *Science*, **184**(4142): 1176–79.

Linnenluecke, M. K. 2013a. 'Can local government drive adaptation to climate change?' *Public Money &Management*, September.

Linnenluecke, M. K. 2013b. *Variations in Decision Makers' Use of Climate Change Information Sources and Impacts on Business Adaptation Choices*', ECCA: European Climate Change Adaptation Conference 2013. Hamburg, Germany: 1st European Climate Change Adaptation Conference (ECCA 2013).

Linnenluecke, M. K., and Griffiths, A. 2010. 'Beyond adaptation: resilience for business in light of climate change and weather extremes.' *Business and Society*, **49**(3): 477–11.

Linnenluecke, M. K., and Griffiths, A. 2011. 'Firm relocation as adaptive response to climate change and weather extremes'. *Global Environmental Change*, **21**(1): 123–33.

Linnenluecke, M. K., and Griffiths, A. 2012. 'Assessing organizational resilience to climate and weather extremes: complexities and methodological pathways'. *Climatic Change*, **113**(3–4): 933–47.

Linnenluecke, M. K., and Griffiths, A. 2013. 'The 2009 Victorian bushfires: a multilevel perspective on organizational risk and resilience'. *Organization & Environment*, **26**(4): 386–411.

Linnenluecke, M. K., Griffiths, A., and Winn, M. 2012. 'Extreme weather events and the critical importance of anticipatory adaptation and organizational resilience in responding to impacts'. *Business Strategy and the Environment*, **21**(1): 17–32.

Lloyd, P. E., and Dicken, P. 1972. *Location in Space: A Theoretical Approach to Economic Geography*. New York: Harper and Row.

Lovins, A. B., Lovins, L. H., and Hawken, P. 1999. 'A road map for natural capitalism'. *Harvard Business Review*, **77**(3): 145.

Malthus, T. R. 1878. *An Essay on the Principle of Population: or, a View of its Past and Present Effects on Human Happiness, with an Inquiry into our Prospects Respecting the Future Removal or Mitigation of the Evils which it Occasions*. London: Reeves and Turner.

Manyena, S. B. 2006. 'The concept of resilience revisited'. *Disasters*, **30**(4): 433–50.

Marsh, G. P. 1864. *Man and Nature: or, Physical Geography as Modified by Human Action*. New York: C. Sribner.

Maugh, T. H. 1979. 'Toxic waste disposal a growing problem'. *Science*, **204**(4395): 819–23.

McGray, H. 2013. 'What do we need to know in a changing climate? A research agenda to support adaptation'. Accessed 24 September 2014, available at: http://www.wri.org/blog/2013/03/what-do-we-need-know-changing-climate-research-agenda-support-adaptation.

McKibbin, W. J., and Wilcoxen, P. J. 2009. 'Uncertainty and climate change policy design'. *Journal of Policy Modeling*, **31**(3): 463–77.

McKinsey and Company 2008. *An Australian Cost Curve for Greenhouse Gas Reduction*. Australia: McKinsey Pacific Rim Inc.

Meadows, D. H., Meadows, D. L., Randers, J., and Behrens, W. W. 1972. *The Limits to Growth: A Report for the Club of Rome's Project on the Predicament of Mankind*. London: Universe Books.

Meehl, G. A., Stocker, T. F., Collins, W. D., Friedlingstein, P., Gaye, A. T., Gregory, J. M., and Co-Authors (eds) 2007. *Climate Change 2007: The Physical Science Basis: Contribution of Working Group I to the Fourth Assessment Report of the Intergovernmental Panel on Climate Change*. Cambridge: Cambridge University Press.

Meinshausen, M., Meinshausen, N., Hare, W., Raper, S. C., Frieler, K., Knutti, R., Frame, D. J., and Allen, M. R. 2009. 'Greenhouse-gas emission targets for limiting global warming to 2°C'. *Nature*, **458**(7242): 1158–62.

Meybeck, A., Lankoski, J., Redfern, S., Azzu, N., and Gitz, V. (eds) 2012. *Building Resilience for Adaptation to Climate Change in the Agriculture Sector: Proceedings of a Joint FAO/OECD Workshop*. Rome, Italy: Food and Agriculture Organization (FAO) and United National Organisation for Economic Co-Operation and Development (OECD).

Michaelis, L. 2003. 'The role of business in sustainable consumption'. *Journal of Cleaner Production*, **11**(8): 915–21.

Mill, J. S. 1848. *Principles of Political Economy with some of their Applications to Social Philosophy*. London: John W. Parker.

Millennium Ecosystem Assessment 2005a. *Ecosystems and Human Well-Being: Synthesis*. Accessed 24 September 2014, available at: http://www.millenniumassessment.org/documents/document.356.aspx.pdf.

Millennium Ecosystem Assessment 2005b. *Millennium Ecosystem Assessment*. New York: United Nations Development Program.

Mills, E. 2005. 'Insurance in a climate of change'. *Science*, **309**(5737): 1040.

Mintzberg, H. 1983. 'The case for corporate social responsibility'. *Journal of Business Strategy*, **4**(2): 3–15.

Mizina, S. V., Smith, J. B., Gossen, E., Spiecker, K. F., and Witkowski, S. L. 1999. 'An evaluation of adaptation options for climate change impacts on agriculture in Kazakhstan'. *Mitigation and Adaptation Strategies for Global Change*, **4**(1): 25–41.

Montgomery, W. D. 2005. *Choice of Policy Measures in Annex B Countries and Impacts on Non-Annex B Countries*. Workshop on Mitigation of Climate Change: Socio-Economic Impacts of Mitigation, Bonn, Germany.

Moser, S. C., and Ekstrom, J. A. 2010. 'A framework to diagnose barriers to climate change adaptation'. *Proceedings of the National Academy of Sciences*, **107**(51): 22026–31.

Moss, R., Edmonds, J. A., Hibbard, K., Manning, M., Rose, S., van Vuuren, D., and Co-authors 2010. 'The next generation of scenarios for climate change research and assessment'. *Nature*, **463**: 747–56.

Munich Re 2007. *Topics Geo: Natural Catastrophes 2006: Analyses, Assessments, Positions*. Munich, Germany: Münchener Rückversicherungs-Gesellschaft.

Munich Re 2009. *Topics Geo: Natural Catastrophes 2008: Analyses, Assessments, Positions*. Munich, Germany: Münchener Rückversicherungs-Gesellschaft.

Munich Re 2012. *Topics Geo: Analyses, Assessments, Positions.* Munich, Germany: Munich, Germany: Münchener Rückversicherungs-Gesellschaft.

Munich Re 2014. *Topics Geo: Natural Catastrophes 2013: Analyses, Assessments, Positions.* Munich, Germany: Münchener Rückversicherungs-Gesellschaft.

Næss, L. O., Bang, G., Eriksen, S., and Vevatne, J. 2005. 'Institutional adaptation to climate change: flood responses at the municipal level in Norway'. *Global Environmental Change*, **15**(2): 125–38.

Nakićenović, N., Alcamo, J., Davis, G., DeVries, B., Fenhann, J., Gaffin, S., and Co-Authors. 2000. *Special Report on Emissions Scenarios: A Special Report of Working Group III of the Intergovernmental Panel on Climate Change.* Cambridge: Cambridge University Press.

National Research Council 2013. *Abrupt Impacts of Climate Change: Anticipating Surprises.* Washington, DC: National Academies Press.

Natural Resources Canada 2014. *Adaptation Platform.* Accessed 24 September 2014, available at: http://www.nrcan.gc.ca/environment/impacts-adaptation/adaptation-platform/10027.

Nepstad, D. C., Stickler, C. M., Soares-Filho, B., and Merry, F. 2008. 'Interactions among Amazon land use, forests and climate: prospects for a near-term forest tipping point'. *Philosophical Transactions of the Royal Society B: Biological Sciences*, **363**(1498): 1737–46.

New, M., Liverman, D., Schroder, H., and Anderson, K. 2011. 'Four degrees and beyond: the potential for a global temperature increase of four degrees and its implications'. *Philosophical Transactions of the Royal Society A: Mathematical, Physical and Engineering Sciences*, **369**(1934): 6–19.

Nicholls, R. J., Marinova, N., Lowe, J. A., Brown, S., Vellinga, P., De Gusmao, D., Hinkel, J., and Tol, R. S. 2011. 'Sea-level rise and its possible impacts given a "beyond 4°C world" in the twenty-first century'. *Philosophical Transactions of the Royal Society A: Mathematical, Physical and Engineering Sciences*, **369**(1934): 161–81.

Nitkin, D., Foster, R., and Medalye, J. 2009. *Business Adaptation to Climate Change: A Systematic Review of the Literature.* Accessed 23 October 2014, available at: http://nbs.net/fr/files/2011/08/NBS_ClimateChange_Concepts_2009.pdf

Noble, I., Huq, S., Anokhin, Y., Carmin, J., Goudou, D., Lansigan, F., and Co-Authors. 2014. 'Adaptation needs and options'. In IPCC (ed.), *Climate Change 2014: Impacts, Adaptation, and Vulnerability: Contribution of Working Group II to the Fifth Assessment Report of the Intergovernmental Panel on Climate Change.* Accessed 24 September 2014, available at: http://ipcc-wg2.gov/AR5/report/.

Nolte, R., Kambrow, C., and Rupp, J. 2011. *ARISCC – Adaptation of Railway Infrastructure to Climate Change*. Berlin: IZT – Institute for Future Studies and Technology Assessment.

Nordhaus, W. D., and Radetzki, M. 1994. *Managing the Global Commons: The Economics of Climate Change*. Cambridge, MA: MIT Press.

Nriagu, J. O., and Pacyna, J. M. 1988. 'Quantitative assessment of worldwide contamination of air, water and soils by trace metals'. *Nature*, **333**(6169): 134–9.

Oberthür, S., and Ott, H. E. 1999. *The Kyoto Protocol: International Climate Policy for the 21st Century*. Berlin and Heidelberg, Germany: Springer.

OECD 2013. *Effective Carbon Prices*. Paris: OECD Publishing.

Oreskes, N. 2004. 'The scientific consensus on climate change'. *Science*, **306**(5702): 1686.

Organisation for Economic Co-operation and Development (OECD) 2012. *Policy Forum on Adaptation to Climate Change in OECD Countries: Summary Note*. Paris: OECD.

Osman-Elasha, B., and Downing, T. 2007. *Lessons Learned in Preparing National Adaptation Programmes of Action in Eastern and Southern Africa*. Oxford: European Capacity Building Initiative.

Ostrom, E. 1999. 'Coping with Tragedies of the Commons'. *Annual Review of Political Science*, 2(1): 493–535.

Ostrom, E. 2009. *Understanding Institutional Diversity*. Princeton: Princeton University Press.

Pachauri, R. K., and Reisinger, A. 2007. *Climate Change 2007: Synthesis Report*. Geneva, Switzerland: IPCC.

Pall, P., Aina, T., Stone, D. A., Stott, P. A., Nozawa, T., Hilberts, A. G., Lohmann, D., and Allen, M. R. 2011. 'Anthropogenic greenhouse gas contribution to flood risk in England and Wales in autumn 2000'. *Nature*, **470**(7334): 382–85.

Pallenbarg, P., van Wissen, L., and van Dijk, J. 2002. 'Firm migration'. In P. McCann (ed.), *Industrial Locaton Economics*. Cheltenham, UK and Northampton, MA, USA: Edward Elgar.

Parfitt, T. 2010. 'Vladimir Putin bans grain exports as drought and wildfires ravage crops'. Accessed 24 September 2014, available at: http://www.theguardian.com/world/2010/aug/05/vladimir-putin-ban-grain-exports. *The Guardian.*

Parry, M. 2009. 'Closing the loop between mitigation, impacts and adaptation'. *Climatic Change*, **96**(1): 23–7.

Parry, M. L., Arnell, N., Berry, P., Dodman, D., Fankhauser, S., Hope, C., and Co-Authors. 2009. *Assessing the Costs of Adaptation to Climate*

Change: A Review of the UNFCCC and other Recent Estimates. London: Imperial College London, Grantham Institute for Climate Change.

Parry, M. L., Canziani, O. F., Palutikof, J.P., and Co-authors 2007. 'Technical summary'. In M. L. Parry, O. F. Canziani, J. P. Palutikof, P. J. van der Linden, and C. E. Hanson (eds), *Climate Change 2007: Impacts, Adaptation and Vulnerability. Contribution of Working Group II to the Fourth Assessment Report of the Intergovernmental Panel on Climate Change*. Cambridge: Cambridge University Press.

Patenaude, G. 2011. 'Climate change diffusion: while the world tips, business schools lag'. *Global Environmental Change*, **21**(1): 259–71.

Pearce, D. W., Barbier, E. B., and Markandya, A. 1989. *Blueprint for a Green Economy*. London: Earthscan Publications.

Perrow, C. 1984. *Normal Accidents: Living with High-Risk Technologies*. New York: Basic Books.

Peters, G. P., Andrew, R. M., Boden, T., Canadell, J. G., Ciais, P., Le Quéré, C., and Co-Authors. 2013. 'The challenge to keep global warming below 2°C'. *Nature Climate Change*, **3**(1): 4–6.

PEW Center on Global Climate Change 2009. *Summary of COP 15 and CMP 5*. Arlington, VA: Pew Center on Global Climate Change.

PEW Center on Global Climate Change 2010. *Summary of COP 16 and CMP 6*. Arlington, VA: Pew Center on Global Climate Change.

PEW Center on Global Climate Change 2011. *Summary of COP 17 and CMP 7*. Arlington, VA: Pew Center on Global Climate Change.

Pineda, C. 2012. *The Cartagena Dialogue: A Bridge to the North-South Divide?* Providence, RI: Brown University, Center for Environmental Studies.

Pinkse, J., and Kolk, A. 2009. *International Business and Global Climate Change*. New York, NY, Abingdon, Oxon: Routledge.

Porter, M., and Reinhardt, F. 2007. 'A strategic approach to climate'. *Harvard Business Review*, **85**(10): 22–3.

Post, J. E., and Altman, B. W. 1994. 'Managing the environmental change process: barriers and opportunities'. *Journal of Organizational Change Management*, **7**(4): 64–81.

PRECIS (Providing Regional Climates for Impacts Studies) 2004. *Generating High Resolution Climate Change Scenarios Using PRECIS*. Accessed 24 September 2014, available at: http://www.metoffice. gov.uk/media/pdf/6/5/PRECIS_Handbook.pdf.

Preston, B. J. 2011. 'The influence of climate change litigation on government and the private sector'. *Climate Law*, **2**(4): 485–513.

Prins, G., and Rayner, S. 2007. 'Time to ditch Kyoto'. *Nature*, 449(7165): 973–75.

Productivity Comission 2012. *Barriers to Effective Climate Change Adaptation: Productivity Commission Inquiry Report*. Canberra: Commonwealth of Australia

Putnam, R. 1993. *Making Democracy Work: Civic Traditions in Modern Italy*. Princeton, NJ: Princeton University Press.

PwC. 2013. *International Threats and Opportunities of Climate Change for the UK*. Accessed 24 September 2014, available at: http://www.pwc.co.uk/sustainability-climate-change/publications/international-threats-and-opportunities-of-climate-change-to-the-uk.jhtml.

Radzicki, M., and Taylor, R. A. 1997. *Introduction to System Dynamics: A Systems Approach to Understanding Complex Policy Issues*. US Department of Energy.

Randalls, S. 2010. 'History of the 2°C climate target'. *Wiley Interdisciplinary Reviews: Climate Change*, **1**(4): 598–605.

Revelle, R., and Suess, H. E. 1957. 'Carbon dioxide exchange between atmosphere and ocean and the question of an increase of atmospheric CO_2 during the past decades'. *Tellus*, **9**(1): 18–27.

Revi, A., Satterthwaite, D., Aragón-Durand, F., Corfee-Morlot, J., Kiunshi, R. B. R., Pelling, M., and Co-Authors. 2014. 'Urban areas'. In IPCC (ed.), *Climate Change 2014: Impacts, Adaptation, and Vulnerability: Contribution of Working Group II to the Fifth Assessment Report of the Intergovernmental Panel on Climate Change*. Accessed 24 September 2014, available at: http://ipcc-wg2.gov/AR5/report/.

Rinaldi, S. M., Peerenboom, J. P., and Kelly, T. K. 2001. 'Identifying, understanding, and analyzing critical infrastructure interdependencies'. *IEEE Control Systems Magazine*, **21**(6): 11–25.

Rockström, J., Steffen, W., Noone, K., Persson, Å., Chapin, F. S., Lambin, E.F., and Co-Authors. 2009a. 'Planetary boundaries: exploring the safe operating space for humanity'. *Ecology and Society*, **14**(2).

Rockström, J., Steffen, W., Noone, K., Persson, Å., Chapin, F. S., Lambin, E. F., and Co-Authors. 2009b. 'A safe operating space for humanity'. *Nature*, **461**(7263): 472–5.

Romo, F., and Schwartz, M. 1995. 'The structural embeddedness of business decisions: the migration of manufacturing plants in New York State, 1960 to 1985'. *American Sociological Review*, **60**: 874–907.

Rooney, S. 2007.' The value of a truly sustainable business strategy'. *Ecos*, 2007(**138**): 27–8.

Sanderson, M., Hemming, D., and Betts, R. 2011. 'Regional temperature and precipitation changes under high-end (≥4°C) global warming'. *Philosophical Transactions of the Royal Society A: Mathematical, Physical and Engineering Sciences*, **369**(1934): 85–98.

Sands, P. 1992. 'The United Nations Framework Convention on Climate Change'. *Review of European Community and International Environmental Law*, **1**(3): 270–7.

Sartori, I., and Hestnes, A. G. 2007. 'Energy use in the life cycle of conventional and low-energy buildings: a review article'. *Energy and Buildings*, **39**(3): 249–57.

Schiermeier, Q. 2010. 'The real holes in climate science'. *Nature*, **463**(7279): 284–7.

Schipper, E. L. F. 2006. 'Conceptual history of adaptation in the UNFCCC process'. *Review of European Community and International Environmental Law*, **15**(1): 82–92.

Schmidheiny, S. 1992. *Changing Course: A Global Business Perspective on Development and the Environment*. Cambridge, MA: MIT Press.

Schneider, S. H. 2004. 'Abrupt non-linear climate change, irreversibility and surprise'. *Global Environmental Change*, **14**(3): 245–58.

Schneider, S. H., Semenov, S., Patwardhan, A., Burton, I., Magadza, C. H. D., Oppenheimer, M., and Co-Authors. 2007. 'Assessing key vulnerabilities and the risk from climate change'. In M. L. Parry, O. F. Canziani, J. P. Palutikof, P. J. van der Linden, and C. E. Hanson (eds), *Climate Change 2007: Impacts, Adaptation and Vulnerability. Contribution of Working Group II to the Fourth Assessment Report of the Intergovernmental Panel on Climate Change*: 779–810. Cambridge, UK: Cambridge University Press.

Schneider, S., Sarukhan, J., Adejuwon, J., Azar, C., Baethgen, W., Hope, C., and Co-Authors. 2001. 'Overview of impacts, adaptation, and vulnerability to climate change'. In J. J. McCarthy, O. F. Canziani, N. A. Leary, D. J. Dokken, and K. S. White (eds), *Climate Change 2001: Impacts, Adaptation, and Vulnerability: Contribution of Working Group II to the Third Assessment Report of the Intergovernmental Panel on Climate Change*. Cambridge: Cambridge University Press.

Schot, J., Brand, E., and Fischer, K. 1997. 'The greening of industry for a sustainable future: building an international research agenda'. *Business Strategy and the Environment*, **6**(3): 153–62.

Schwartz, P., and Randall, D. 2003. *An Abrupt Climate Change Scenario and its Implications for United States National Security*. Washington, DC: Department of Defense.

Senge, P. M., and Carstedt, G. 2001. 'Innovating our way to the next industrial revolution'. *MIT Sloan Management Review*, **42**(2): 24–38.

Shardul, A., and Samuel, F. (eds). 2008. *Economic Aspects of Adaptation to Climate Change Costs, Benefits and Policy Instruments: Costs, Benefits and Policy Instruments*. Paris: OECD Publishing.

Sharma, S., and Aragón-Correa, J. A. 2005. 'Corporate environmental strategy and competitive advantage: a review from the past to the future'. In S. Sharma, and J. A. Aragón-Correa (eds), *Corporate Environmental Strategy and Competitive Advantage*. Cheltenham, UK and Northampton, MA, USA: Edward Elgar.

Sherwood, S. C., and Huber, M. 2010. 'An adaptability limit to climate change due to heat stress'. *Proceedings of the National Academy of Sciences*, **107**(21): 9552–55.

Shrivastava, P. 1994. 'Castrated environment: greening organizational studies'. *Organization Studies*, **15**(5): 705–26.

Shrivastava, P., and Kennelly, J. J. 2013. 'Sustainability and place-based enterprise'. *Organization & Environment*, **26**(1): 83–101.

SimCLIM 2013. *SimCLIM Software Package*. Accessed 24 September 2014, available at: http://www.climsystems.com/simclim/.

Smit, B., and Wandel, J. 2006. 'Adaptation, adaptive capacity and vulnerability'. *Global Environmental Change*, **16**(3): 282–92.

Smit, B., Burton, I., Klein, R. J., and Street, R. 1999. 'The science of adaptation: a framework for assessment'. *Mitigation and Adaptation Strategies for Global Change*, **4**(3–4): 199–213.

Smit, B., Burton, I., Klein, R., and Wandel, J. 2000. 'An anatomy of adaptation to climate change and variability'. *Climatic Change*, **45**: 223–51.

Smith, B., Burton, I., Klein, R. J., and Wandel, J. 2000. 'An anatomy of adaptation to climate change and variability'. *Climatic Change*, **45**(1): 223–51.

Smith, P., Martino, D., Cai, Z., Gwary, D., Janzen, H., Kumar, P., and Co-authors. 2007. 'Agriculture'. In B. Metz, P. Bosck, R. Dave, and L. Meyer (eds), *Climate Change 2007: Mitigation of Climate Change. Contribution of Working Group III to the Fourth Assessment Report of the Intergovernmental Panel on Climate Change*. Cambridge and New York: Cambridge University Press.

Solomon, S., Qin, D., Manning, M., Alley, R. B., Berntsen, T., Bindoff, N. L., and Co-Authors. 2007a. 'Technical Summary'. In S. Solomon, D. Qin, M. Manning, Z. Chen, M. Marquis, K. B. Averyt, M. Tignor, and H. L. Miller (eds), *Climate Change 2007: The Physical Science Basis. Contribution of Working Group I to the Fourth Assessment Report of the Intergovernmental Panel on Climate Change*. Cambridge and New York: Cambridge University Press.

Solomon, S., Qin, D., Manning, M., Chen, Z., Marquis, M., Averyt, K. B., and Co-Authors (eds) 2007b. *Climate Change 2007: The Physical Science Basis. Contribution of Working Group I to the Fourth Assessment Report of the Intergovernmental Panel on Climate Change*. Cambridge: Cambridge University Press.

Starik, M. 1995. 'Should trees have managerial standing? Toward stakeholder status for non-human nature'. *Journal of Business Ethics*, **14**(3): 207–17.

Starik, M., and Marcus, A. A. 2000. 'Introduction to the special research forum on the management of organizations in the natural environment: a field emerging from multiple paths, with many challenges ahead'. *Academy of Management Journal*, **43**(4): 539–46.

Steffen, W. 2013. *The Angry Summer.* Canberra: Climate Commission.

Stern, N. 2007. *The Economics of Climate Change: The Stern Review.* Cambridge: Cambridge University Press.

Stern, N., Peters, S., Bakhshi, V., Bowen, A., Cameron, C., Catovsky, S., and Co-Authors. 2006. *Stern Review: The Economics of Climate Change.* London: HM Treasury.

Stott, P. A., Stone, D. A., and Allen, M. R. 2004. 'Human contribution to the European heatwave of 2003'. *Nature*, **432**(7017): 610–14.

Streets, D. G., and Glantz, M. H. 2000. 'Exploring the concept of climate surprise'. *Global Environmental Change*, **10**(2): 97–107.

Sussman, F. G., and Freed, J. R. 2008. *Adapting to Climate Change: A Business Approach.* Arlington, VA: Pew Center on Global Climate Change.

Sutter, C., and Parreño, J. C. 2007. 'Does the current Clean Development Mechanism (CDM) deliver its sustainable development claim? An analysis of officially registered CDM projects'. *Climatic Change*, **84**(1): 75–90.

Thaler, R. 1999. 'Mental accounting matters'. *Behavioral Descision Making*, **12**(3): 183–206.

Tol, R. S. 2005. 'Adaptation and mitigation: trade-offs in substance and methods'. *Environmental Science and Policy*, **8**(6): 572–78.

Trenberth, K. 2005. 'Uncertainty in hurricanes and global warming'. *Science*, **308**(5729): 1753–54.

Trenberth, K. 2010. 'More knowledge, less certainty'. *Nature Reports Climate Change*, **4**(February): 20–1.

Trenberth, K. E., and Shea, D. J. 2006. 'Atlantic hurricanes and natural variability in 2005'. *Geophysical Research Letters*, **33**(12): L12704.

Trenberth, K., Jones, P., Ambenje, P., Bojariu, B., Easterling, D., and Co-authors 2007. 'Observations: surface and atmospheric climate change'. In S. Solomon, Qin, D, M. Manning, Z. Chen, M. Marquis, K. Averyt, M. Tignor, and H. Miller (eds), *Climate Change 2007: The Physical Science Basis. Contribution of Working Group I to the Fourth Assessment Report of the Intergovernmental Panel on Climate Change.* Cambridge and New York: Cambridge University Press.

UKCIP 2004. *Costing the Impacts of Climate Change in the UK: Overview of Guidelines.* Accessed 24 September 2014, available at:

http://www.ukcip.org.uk/wordpress/wp-content/PDFs/Costings_overview. pdf.

UNEP 2012. *Business and Climate Change Adaptation: Toward Resilient Companies and Communities.* Accessed 24 September 2014, available at: http://unglobalcompact.org/docs/issues_doc/Environment/climate/ Business_and_Climate_Change_Adaptation.pdf.

UNEP 2013. *GEO-5 for Business: Impacts of a Changing Environment on the Corporate Sector.* Accessed 24 September 2014, available at: http://www.unep.org/geo/pdfs/geo5/geo5_for_business.pdf.

UNFCCC 2002. *Annotated Guidelines for the Preparation of National Adaptation Programmes of Action.* Accessed 24 September 2014, available at: https://unfccc.int/files/cooperation_and_support/ldc/ application/pdf/annguide.pdf.

UNFCCC 2007a. *Climate Change: Impacts, Vulnerabilities and Adaptation in Developing Countries.* Bonn, Germany: Climate Change Secretariat (UNFCCC).

UNFCCC 2007b. *Uniting on Climate: A Guide to the Climate Change Convention and the Kyoto Protocol.* Bonn, Germany: Climate Change Secretariat (UNFCCC).

UNFCCC 2009. *The Need for Mitigation.* Accessed 24 September 2014, available at: https://unfccc.int/files/press/backgrounders/application/ pdf/press_factsh_mitigation.pdf.

UNFCCC 2010. *Adaptation Assessment, Planning and Practice: An Overview of the Nairobi Work Programme on Impacts, Vulnerability and Adaptation to Climate Change.* Bonn, Germany: Climate Change Secretariat (UNFCCC).

UNFCCC 2011a. *Assessing the Costs and Benefits of Adaptation Options: An Overview of Approaches in The Nairobi Work Programme on Impacts, Vulnerability and Adaptation to Climate Change.* Accessed 24 September 2014, available at: http://unfccc.int/files/adaptation/nairobi_work_ programme/knowledge_resources_and_publications/application/pdf/2011_ nwp_costs_benefits_adaptation.pdf.

UNFCCC 2011b. *Draft decision [-/CP.16] of the Ad Hoc Working Group on Long-term Cooperative Action under the Convention.* Accessed 24 September 2014, available at: http://unfccc.int/resource/docs/2010/ awglca13/eng/l07.pdf.

UNFCCC 2011c. *Draft decision [-/CP.17]: Outcome of the work of the Ad Hoc Working Group on Long-term Cooperative Action under the Convention.* Accessed 24 September 2014, available at: http://unfccc. int/files/meetings/durban_nov_2011/decisions/application/pdf/cop17_lca outcome.pdf.

UNFCCC 2013a. *Background on the UNFCCC: The International Response to Climate Change.* Accessed 24 September 2014, available at: http://unfccc.int/essential_background/items/6031.php.

UNFCCC 2013b. *Bali Climate Change Conference – December 2007.* Accessed 24 September 2014, available at: http://unfccc.int/meetings/bali_dec_2007/meeting/6319.php.

UNFCCC 2013c. *Cancun Adaptation Framework: Chronology.* Accessed 24 September 2014, available at: http://unfccc.int/adaptation/items/6052.php.

UNFCCC 2013d. *Cancun Climate Change Conference – November 2010.* Accessed 24 September 2014, available at: http://unfccc.int/meetings/cancun_nov_2010/meeting/6266.php.

UNFCCC 2013e. *Copenhagen Climate Change Conference – December 2009.* Accessed 24 September 2014, available at: http://unfccc.int/meetings/copenhagen_dec_2009/meeting/6295.php.

UNFCCC 2013f. *COSMIC2 (Country Specific Model for Intertemporal Climate Vers. 2).* Accessed 24 September 2014, available at: https://unfccc.int/adaptation/nairobi_work_programme/knowledge_resources_and_publications/items/5346.php.

UNFCCC 2013g. *Doha Climate Change Conference – November 2012.* Accessed 24 September 2014, available at: https://unfccc.int/meetings/doha_nov_2012/meeting/6815.php.

UNFCCC 2013h. *Making Those First Steps Count: An Introduction to the Kyoto Protocol.* Accessed 24 September 2014, available at: https://unfccc.int/essential_background/kyoto_protocol/items/6034.php.

UNFCCC 2013i. *The Mechanisms under the Kyoto Protocol: Emissions Trading, the Clean Development Mechanism and Joint Implementation.* Accessed 24 September 2014, available at: http://unfccc.int/kyoto_protocol/mechanisms/items/1673.php.

UNFCCC 2013j. *Parties and Observers.* Accessed 24 September 2014, available at: http://unfccc.int/parties_and_observers/items/2704.php.

UNFCCC 2014. *Adaptation Private Sector Initiative (PSI).* Accessed 24 September 2014, available at: http://unfccc.int/adaptation/workstreams/nairobi_work_programme/items/4623.php.

UNFCCC n.d. *Adaptation under the Frameworks of the CBD, the UNCCD and the UNFCCC.* Accessed 24 September 2014, available at: http://unfccc.int/resource/docs/publications/adaptation_eng.pdf.

UNFCCC Secretariat 2005. *Compendium on Methods and Tools to Evaluate Impacts of, and Vulnerability and Adaptation to, Climate Change.* Accessed 24 September 2014, available at: https://unfccc.int/files/adaptation/methodologies_for/vulnerability_and_adaptation/application/pdf/200502_compendium_methods_tools_2005.pdf

UNFCCC Secretariat n.d. *The Nairobi Work Programme: An Overview*. Accessed 24 September 2014, available at: http://unfccc.int/files/adaptation/application/pdf/nwpleaflet_overview.pdf.

United Nations 1992. *United Nations Framework Convention on Climate Change*. Accessed 24 September 2014, available at: http://unfccc.int/resource/docs/convkp/conveng.pdf.

United Nations 2000. *United Nations Millennium Declaration*. (A/res/55/2).

United Nations 2012. *Resilient people, resilient planet: A future worth choosing*. New York: United Nations.

United Nations Conference of Parties 2007. *Bali Action Plan. UN Framework Convention on Climate Change's 13th Conference of the Parties (COP 13)*. Accessed 24 September 2014, available at: http://unfccc.int/files/meetings/cop_13/application/pdf/cp_bali_action.pdf.

United Nations Global Compact and United Nations Environment Programme 2012. *Business and climate change adaptation: toward resilient companies and communities*. New York: UN Global Compact Office.

US Global Change Research Act 1990. *Public Law 101-606(11/16/90) 104 Stat. 3096-3104, approved November 16, 1990 (updated February 6, 2004)*. Accessed 23 October 2014, available at: www.gcrio.org/gcact1990.html.

US Global Change Research Program 2013. *National Climate Assessment*. Accessed 24 September 2014, available at: http://www.globalchange.gov/what-we-do/assessment.

US Global Change Research Program 2013. *Our Changing Planet: A supplement to the President's Budget for Fiscal Year 2013*. Washington, DC.

US Senate 1997. *Byrd Hagel Resolution. 105th Congress, First Session*. Accessed 24 September 2014, available at: http://www.gpo.gov/fdsys/pkg/BILLS-105sres98rs/pdf/BILLS-105sres98rs.pdf.

Uzzi, B. 1996. The sources and consequences of embeddedness for the economic performance of organizations: the network effect. *American Sociological Review*, **61**: 674–98.

van Vuuren, D., Edmonds, J., Kainuma, M., Riahi, K., and Co-authors. 2011. 'The representative concentration pathways: an overview'. *Climatic Change*, **109**: 5–31.

Verschuuren, J. (ed.) 2013. *Research Handbook on Climate Change Adaptation Law*. Cheltenham, UK and Northhampton, MA, USA: Edward Elgar.

Verschuuren, J. 2010. 'Climate change: rethinking restoration in the European Union's Birds and Habitats Directives'. *Ecological Restoration*, **28**(4): 431–39.

Vince, C. A., Nordstrom, P. E., and Speed-Bost, R. Y. 2007. 'New Orleans calamity gives lessons in utility preparedness'. *Natural Gas and Electricity*, **23**: 1–31.

Vincent, K., Naess, L., and Goulden, M. 2013. 'National level policies versus local level realities – can the two be reconciled to promote sustainable adaptation?' In L. Sygna, K. O'Brien, and J. Wolf (eds), *A Changing Environment for Human Security: Transformative Approaches to Research, Policy, and Action*. London: Earthscan, Routledge.

Vis, M., Klijn, F., De Bruijn, K., and Van Buuren, M. 2003. 'Resilience strategies for flood risk management in the Netherlands'. *International Journal of River Basin Management*, **1**(1): 33–40.

Walley, N., and Whitehead, B. 1994. 'It's not easy being green'. *Harvard Business Review*, **72**(3): 46–52.

Warner, K., Ehrhart, C., de Sherbinin, A., Adamo, S., and Chai-Onn, T. 2009. *In Search of Shelter: Mapping the Effects of Climate Change on Human Migration and Displacement*. Accessed 24 September 2014, available at: http://www.careclimatechange.org/.

WCED (World Commission on Environment and Development) 1987. *Our Common Future*. Oxford: Oxford University Press.

Weick, K. E., Sutcliffe, K. M., and Obstfeld, D. 2005. 'Organizing and the process of sensemaking'. *Organization Science*, **16**(4): 409–21.

Weiskel, T. C. 1989. 'The ecological lessons of the past: an anthropology of environmental decline'. *The Ecologist*, **19**(3): 98–103.

Whiteman, G., and Cooper, W. 2011. 'Ecological sensemaking'. *Academy of Management Journal*, **54**: 889–911.

Whiteman, G., Walker, B., and Perego, P. 2013. 'Planetary boundaries: ecological foundations for corporate sustainability'. *Journal of Management Studies*, **50**(2): 307–36.

Wigley, T. M. L., Raper, S. C. B., Hulme, M., and Smith, S. 2000. The MAGICC/SCENGEN Climate Scenario Generator: Version 2.4. Norwich: Climatic Research Unit, University of East Anglia.

Wilbanks, T. J. 2003. 'Integrating climate change and sustainable development in a place-based context'. *Climate Policy*, **3**(Supplement 1): S147–54.

Wilbanks, T. J., and Sathaye, J. 2007. 'Integrating mitigation and adaptation as responses to climate change: a synthesis'. *Mitigation and Adaptation Strategies for Global Change*, **12**(5): 957–62.

Wilbanks, T. J., Leiby, P., Perlack, R., Ensminger, J. T., and Wright, S. B. 2007a. 'Toward an integrated analysis of mitigation and adaptation: some preliminary findings'. *Mitigation and Adaptation Strategies for Global Change*, **12**(5): 713–25.

Wilbanks, T. J., Romero Lankao, P. Bao, M., Berkhout, F., Cairncross, S., Ceron, J.-P., and Co-Authors. 2007b. 'Industry, settlement and society'. In M. L. Parry, O. F. Canziani, J. P. Palutikof, P. J. van der Linden, and C. E. Hanson (eds), *Climate Change 2007: Impacts, Adaptation and Vulnerability. Contribution of Working Group II to the Fourth Assessment Report of the Intergovernmental Panel on Climate Change.* Cambridge: Cambridge University Press.

Willows, R., and Connell, R. 2003. *Climate Adaptation: Risk, Uncertainty and Decision-making, UKCIP Technical Report.* Oxford: UKCIP.

Winn, M. I., and Kirchgeorg, M. 2005. 'The siesta is over: a rude awakening from sustainability myopia'. In S. Sharma, and J. A. Aragón-Correa (eds), *Corporate Environmental Strategy and Competitive Advantage.* Cheltenham, UK and Northhampton, MA, USA: Edward Elgar.

Winn, M. I., Kirchgeorg, M., Griffiths, A., Linnenluecke, M. K., and Gunther, E. 2011. 'Impacts from climate change on organizations: a conceptual foundation'. *Business Strategy and the Environment*, **20**(3): 157–73.

Woods, D. 2006. 'Essential characteristics of resilience'. *Resilience Engineering: Concepts and Precepts*: 21–34: Aldershot: Ashgate.

World Bank 2013a. *Mapping Carbon Pricing Initiatives: Developments and Prospects.* Washington, DC: Ecofys and World Bank.

World Bank 2013b. *Turn Down the Heat: Climate Extremes, Regional Impacts, and the Case for Resilience.* Accessed 24 September 2014, available at: http://www-wds.worldbank.org/external/default/WDSContent Server/WDSP/IB/2013/06/14/000445729_20130614145941/Rendered/PDF/ 784240WP0Full00D0CONF0to0June19090L.pdf.

World Economic Forum 2013. *Global Risks 2013* (8th edn). Cologny/ Geneva, Switzerland: World Economic Forum.

World Meteorological Organization 2014. *Climate Information.* Accessed 24 September 2014, available at: http://www.wmo.int/pages/themes/ climate/index_en.php#.

World Resources Institute 2009. *Making Climate Your Business: Private Sector Adaptation in Southeast Asia.* Accessed 24 September 2014, available at: http://www.wri.org/sites/default/files/pdf/making_climate_ your_business.pdf.

World Resources Institute 2010. *Reflections on the Cancun Agreements.* Accessed 24 September 2014, available at: http://www.wri.org/stories/ 2010/12/reflections-cancun-agreements.

World Resources Institute n.d. *Mainstreaming Climate Change Adaptation: The Need and Role of Civil Society Organisations.* Accessed 24 September 2014, available at: http://www.wri.org/node/40292.

Wyss, R., Luthe, T., and Abegg, B. 2014. 'Building resilience to climate change – the role of cooperation in alpine tourism networks'. *Local Environment* (pending publication): 1–15.

Yohe, G., and Tol, R. 2002. 'Indicators for social and economic coping capacity – moving towards a working definition of adaptive capacity'. *Global Environmental Change*, **12**(1): 25–40.

Index